Carl Fedtke

Biochemistry and Physiology of

Herbicide Action

With 43 Figures and 58 Tables

Springer-Verlag
Berlin Heidelberg New York 1982

Dr. Carl Fedtke

D-5090 Leverkusen, Bayerwerk
Federal Republic of Germany

ISBN 3-540-11231-6 Springer-Verlag Berlin Heidelberg New York
ISBN 0-387-11231-6 Springer-Verlag New York Heidelberg Berlin

Library of Congress Cataloging in Publication Data
Fedtke, Carl, 1940–. Biochemistry and physiology of herbicide action
Bibliography: p. Includes index.
1. Plants, Effect of herbicides on. 2. Plants – Metabolism. I. Title.
QK753.H45F4. 581.1'33. 81-21357
ISBN 0-387-11231-6 (U.S.) AACR2

Typesetting, printing, and binding: Brühlsche Universitätsdruckerei, Giessen
2152/3140-543210

Foreword

Herbicides are part of modern agricultural production systems and therefore contribute significantly to the economy of agricultural products. At the same time, herbicides are potent and specific inhibitors of plant metabolism and may therefore be used as valuable tools in basic plant physiological research. A well-known example is the photosynthesis-inhibiting herbicide diuron, known to plant physiologists as DCMU, which has become one of the essentials in modern photosynthesis research. Similarly, knowledge in other areas of plant metabolism may be advanced by the use of herbicides as specific inhibitors.

This book describes the effects of herbicides on the metabolism of higher plants from the viewpoint of the plant physiologist. The material of this book is therefore, as far as possible, divided into areas of metabolism. This book intends (1) to present the reader with current knowledge and views in the area of herbicide modes of action and (2) to promote the future use of herbicides as metabolic inhibitors in plant physiological research to the advantage of both, the pesticide and the plant sciences.

I wish to express my thanks to my colleagues and friends Prof. N. Amrhein, Prof. E. Elstner, Dr. L. Eue, Dr. J. Konze, Dr. K. Lürssen, Dr. W. Oettmeier, Dr. H. Quader, Dr. R. R. Schmidt, Dr. R. H. Shimabukuro, Dr. J. Stetter, Prof. Dr. A. Trebst, and Prof. Dr. R. Wegler for the encouragement and help during the preparation of the manuscript, and to Mr. J. Hens for the drawings. Though I owe them my sincere thanks, any shortcomings, misinterpretations, and faults are of course in my own responsibility.

Leverkusen, January 1982 — Carl Fedtke

Table of Contents

A. Approaches to and Definitions of the Mechanisms of Action of Herbicides

I. Applications, Morphological, Physiological, and Biochemical Mechanisms of Herbicides

When raising the question of the mechanism of the action of herbicides, the type of approach and the type of answer will differ considerably depending on the viewpoint of the investigator. An answer to a certain pre-emergence herbicide might be that the compound is taken up by the imbibing seed and severely inhibits root growth of the emerging seedling (e.g. p. 148); for a post-emergence herbicide the answer might be that it rapidly dessicates established leaf tissue, but only in the light (e.g. p. 89). These answers use visual information. However, a vast amount of information has been accumulated over the recent years on the cytological, physiological, biochemical, and molecular aspects of herbicide action on different plant systems, thus enabling us to answer the question of the mechanism of this action much more precisely.

In order to classify the available information on the mechanism of action of herbicides, six different types of data can be distinguished (Table 1 a). This classification may also serve as a guideline in the elucidation of the mechanism of action of new herbicides. Information on the application technique (a) and on sensitive tissues (b) will always be available. This information allows to classify a compound as "germination inhibitor" (p. 148) "contact herbicide" (p. 92), "bleaching herbicide" (p. 105), "desiccant" (p. 89), or "hormone killer" (p. 172). The cytological viewpoint (c) has been explored to a rather minor extent. Data on the accumulation of radioactively labelled herbicides may be obtained either by microautoradiography or subcellular fractionation of a tissue homogenate. However, preferential accumulation of a herbicide in one subcellular structure, e.g. the chloroplast (p. 173), does not necessarily mean that this organelle is or contains the site of action. Accordingly, the first manifestation of visual damage in a subcellular structure, e.g. the chloroplast membrane (p. 182), does not necessarily imply that this organelle is spatially close to the site of action. Therefore, data on the cytological action, interesting and valuable as they might be in relation to other results on herbicidal interferences, did not prove to be very useful in finding the primary site of action.

Information on the physiology (d) and biochemistry (e) of herbicide action cannot always be clearly assigned to one of these two areas. Physiological effects are often ment to be sublethal effects of the herbicide, which means indirect changes of the growth habit and composition of the plant (p. 70). In this context, however, direct herbicidal influences on the physiological aspects of plant metabolism are considered, and "physiological" is understood here as the interplay of the individual biochemical reactions and pathways as they contribute to plant metabolism, growth, and reproduction.

The influence of herbicides on the physiology of the plant (d) is mostly measured as an influence on in vivo metabolic systems, e.g. photosynthetic gas exchange with leaf pieces or protein synthesis by radioactive amino acid incorporation. The inhibition by a herbicide of a metabolic pathway, e.g. protein synthesis, of course does not necessarily imply a primary herbicidal action on this pathway; the measured inhibition might be the consequence of another metabolic block on which the investigated pathway depends (e.g. p. 120). Only in vitro investigations with the isolated enzyme system give reliable data on the primary herbicidal influence on a metabolic pathway. These may be termed biochemical investigations (e). Unfortunately, many metabolic pathways in plants at present may not or only insufficiently be studied by biochemical methods because the appropriate in vitro methods are unavailable, or, even worse, because the primarily inhibited enzyme reactions are not yet known. Suitable examples for well-known biochemical systems are the isolated chloroplasts (p. 21) and mitochondria (p. 114).

The in vitro inhibition of a biochemical system must of course be seen as a function of inhibitor concentration (p. 7), and the reliability of the in vitro inhibition for in vivo conditions, especially in the natural field habitat, must always be clearly documented.

The second type of valuable biochemical information comes from plant composition studies. Changes in the concentration of plant metabolites or constituents after herbicidal application often allow to conclude what has originally happened. The crucial question then is: what did happen first, what second etc. (p. 11)? From a drop in the level of soluble reducing sugars it may be inferred that photosynthesis is inhibited (p. 72); a rise in the concentration of malonedialdehyde should originate from the occurrence of activated oxygen generated by the action of a herbicide (p. 69). In each case, alternative explanations may be proposed; one way to clarify the situation is to carefully analyze the kinetics of the biochemical changes. Another way would be to use structure analogues of the herbicide which are not phytotoxic.

A "true" biochemist will always ask the question of the "exact" site of action, which means the site of molecular interaction. Whereas the physiological answer to the question of the mechanism of action aims at a metabolic pathway (photosynthesis p. 19, respiration p. 115, carotenoid biosynthesis p. 99, mitosis p. 123 etc.), the biochemical answer tries to define the enzyme or enzyme complex which is inhibited (e.g. "B-protein" in chloroplast membranes p. 58, reducing side of pigment system I in chloroplast membranes p. 22, enzymes in carotenoid biosynthesis p. 103). The ultimate answer will be a description of the molecular site of herbicide action (f). This will usually be done by studying the binding of a compound to a relevant enzyme or protein fraction, by demonstrating the chemical structure correlations, by determining the binding site constants, and by correlating binding with the inhibitory or regulatory activity (p. 51–59). The idea of the binding site implies that the herbicide, in order to exert its phytotoxic action, will bind to a specific site on some protein(s) in the plant. When following current ideas on enzyme-sub-

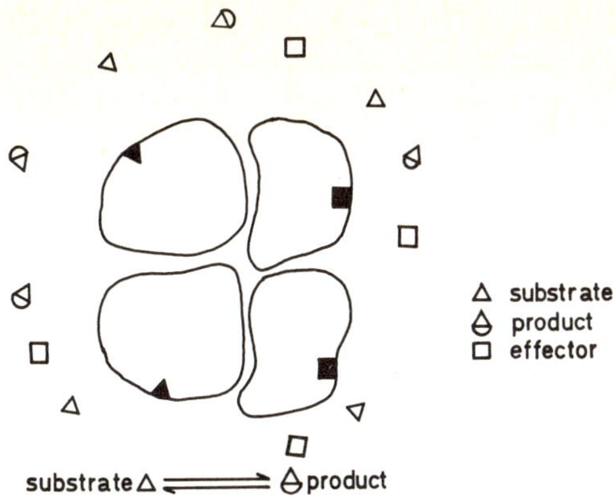

Fig. 1. Binding sites and catalytic activity of a regulatory enzyme protein

strate interactions, we anticipate a specialized location on the enzyme protein which specifically binds the substrate, then turns it into the product and vice versa (Fig. 1). The enzyme protein might further be a regulatory protein comprising two sites, a "substrate site" and a "regulatory site". The regulatory site controls the activity of the substrate site, allowing an effector to regulate the enzyme activity by binding to the regulatory site. More than one effector site for both positive and negative control are possible. According to this simplified scheme the herbicidal compound might bind to a substrate site or to a regulatory site. Herbicides binding to the substrate site should to some extent mimic the true substrate, thereby "fooling" the enzyme protein. This is e.g. true for *glyphosate* (p. 185). The case of the auxin herbicides (pp. 161 and 167) is different: the auxin-binding protein contains only an effector site because it does not chemically alter the bound "substrate", in this case *2,4-D* or *IAA*. But by binding the auxin molecule the protein, possibly via a conformational change, becomes the trigger of the well-known auxin responses. In the case of the "32 KD-protein", which binds the herbicides which inhibit at the reducing side of photosystem II, the natural function of the protein and the possible native "substrate" or "effector" are just emerging (p. 59).

One of the reasons for carrying out studies on the mechanism of herbicide action certainly is to understand why and how a susceptible plant is damaged by a herbicide. To answer this question on a molecular level has often proved to be a very difficult task. In the case of rapid action a toxic molecule should be implied which often is not the herbicide itself but is produced by the plant under the herbicidal influence. In many cases where interference with the photosynthetic pigment system of the plant is involved, activated oxygen species appear to be the toxic agent (pp. 60, 68, 89, 94, 105). In the case of slow action without immediate toxic symptoms but with abnormal growth characteristics the loss of an important metabolic system (e.g. microtubules, p. 128) or the loss of normal growth regulation (e.g. auxin herbicides, p. 172) may provide a guideline for the understanding of herbicide action.

We group herbicides according to related structures and/or mechanisms of action. This is a necessary but often disturbing way. Because each chemical structure represents an individual compound, each should basically be considered individually. Well-known examples show that for a series of compounds with the same site of action, individuals exist which, in addition to the common site, inhibit at a second different site (e.g. *fluometuron*, inhibiting photosynthesis and carotenoid biosynthesis). Generally, depending on concentration, each chemical may inhibit a large number of metabolic pathways, and the mechanism of action in vivo might include the action at more than one site.

One final note should be given to the use of the terms mechanism, mode, or site of action. Whereas "mechanism of action" or "mode of action" will refer to the whole complex of herbicide-plant interferences, "site of action" specifically means the sensitive enzyme-protein, metabolic step, or binding site. The main sites of action which are currently known for existing herbicides appear to belong to the endomembrane system of the plant cell. Whether this is more than a current incidence remains to be seen.

II. Basic, Intermediate, and Secondary Plant Metabolism

In order to better understand the value of data from physiological or biochemical systems, it is helpful to classify plant metabolism into "basic", "intermediate", and "secondary" metabolism (Fig. 2). Basic metabolism includes the production of organic carbon compounds by photosynthesis, the generation of high-energy chemi-

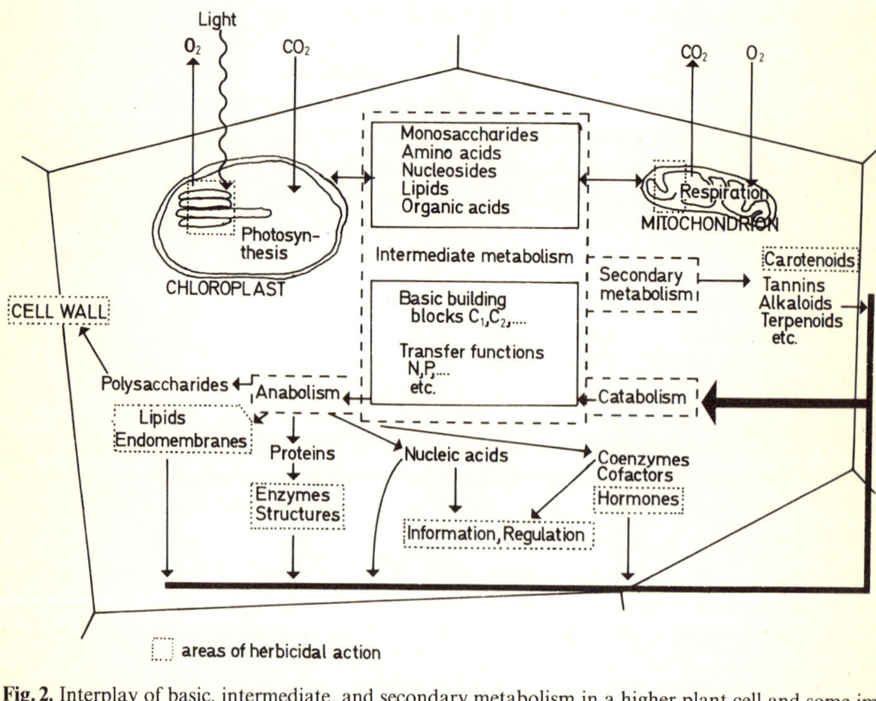

Fig. 2. Interplay of basic, intermediate, and secondary metabolism in a higher plant cell and some important areas of herbicide action

cal bonds through respiration and oxidative phosphorylation (in the form of ATP), and the synthesis of the basic cellular polymers such as proteins, nucleic acids, starch, and cellulose. Via the intermediate metabolism all degrading and synthetic pathways are connected; this comprises the pools of small organic molecules (C-1, C-2, etc.) generated by degradation and/or necessary for new synthesis, their interconversion and incorporation into new building blocks. In the secondary metabolism innumerable different and specific plant compounds like alkaloids, pectins, lignin, coumarins, phenylpropanes, anthocyanins, growth hormones, tannins, etc. are synthesized.

Herbicides may interfere with any one of these pathways which contribute to the metabolism and growth of plants. Experience has shown that metabolic pathways specific for plant tissues contain most of the known sites of action, e.g. photosynthesis, carotenoid synthesis, specific plant regulatory systems like auxin hormones, etc. (Fig. 2). In general, herbicides are therefore relatively nontoxic to animals including mammals. The metabolic pathways specific for plant tissues are found in the basic and secondary plant metabolism, while the intermediary metabolism is largely identical for all living nucleate cells.

III. Supercomplex, Complex, and Defined Systems

How can we obtain the data necessary for the elucidation of the mechanism of herbicide action as listed in Table 1 a? Of course, this is done by experimentation with either intact plants, plant tissue pieces or preparations obtained from disrupted plant tissue (in vitro studies). The potential amount of metabolism revealed by either intact plants, plant tissue pieces or preparations from disrupted plant tissue will decrease in the given order. It may therefore be allowed to classify the respective systems used in studies on the mechanism of herbicide action as "supercomplex", "complex", or "defined systems" (Table 1 b).

"Supercomplex systems", generally represented by intact plants and measuring growth by dry weight increase or another growth-related parameter like e.g. shoot length, might already produce specific valuable information. If combined with

Table 1 a. Information on the mode of action of herbicides as obtained from different types of observations and test systems

Information source	Information type
a. Application technique	Sensitive growth phase or organ
b. Damaged tissue	Sensitive cell type: meristematic, elongating, assimilating, etc.
c. Cytological and microscopical observations	Cellular and subcellular actions
d. Physiological experimentation	In vivo inhibited metabolic pathways
e. Biochemical experimentation	In vitro inhibited metabolic pathways, effects on biochemical composition
f. Protein/enzyme interactions with herbicides	Sensitive enzymatic reaction and/or binding protein

Table 1b. Classification of plant tissues, plant tissue preparations and methods of investigation used in the search for the mechanism of herbicide action

Classification	Data relevance	Type of experimentation
Supercomplex systems	Growth related	Anatomy, dry weight, morphology, etc.
Complex systems	Metabolism related	Tracer studies, gas exchanges, estimation of cell components, etc.
Defined systems	Site directed	In vitro studies with enzymes or organelles, binding studies, etc.

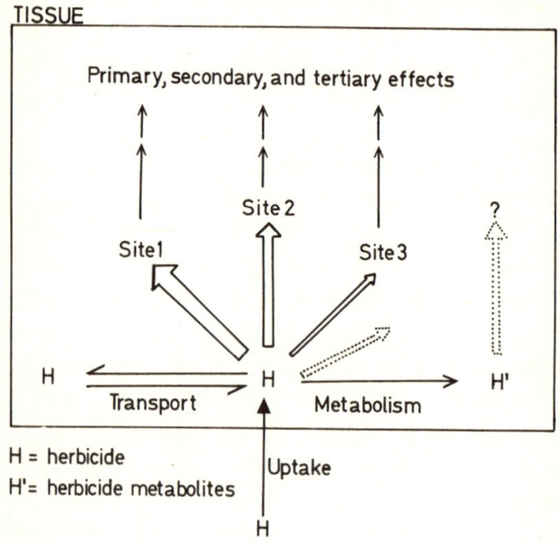

Fig. 3. Model showing herbicide uptake, transport, metabolism, and multiple inhibitory actions in intact plants

morphological and anatomical studies, a specific site of action might already be assumed, like e.g. for compounds inhibiting mitosis and producing swollen root tissue in grass seedlings (p. 129). Generally, additional systems aimed at specific metabolic pathways will be employed. These will usually still be "complex systems" where not only the selected metabolic pathway proceeds. The problem might partially be overcome by e.g. feeding specific radioactive precursors in tracer studies to follow a synthetic pathway (e.g. ^{14}C-leucine→protein, ^{3}H-thymidine→DNA, ^{3}H-uridine→RNA). In case the selected pathway is quantitatively prominent like the carbon dioxide/oxygen gas exchange in respiration (dark measurement) and photosynthesis (light measurement), the interpretation is quite easy. However, when measuring the oxygen uptake, other pathways in addition to respiration have to be considered (p. 60).

In "complex" or "supercomplex systems" involving more than one metabolic pathway, the possibility of the occurrence of more than one site of action has to be considered (Fig. 3) (e.g. pp. 31, 108, *fluometuron*). Since a given herbicide concentration in the tissue, when acting at more than one site, will inhibit these to a different extent, a complicated situation arises with several interacting primary, secondary, and tertiary effects.

Finally, in vitro studies may usually be classified as "defined systems" in that only the investigated pathway proceeds. However, problems arising here include the relevance of the data for in vivo conditions or even the possibility of handling "artifact systems" totally unrelated to the intact plant. A further possible approach related to defined systems is through the use of model systems (e.g. photosynthesis inhibitors, p. 53).

IV. Kinetics and Dose-Response Curves

It might be expected that an inhibition occurs rapidly and at low concentrations when reflecting the primary site of action of a herbicide. An example is given in Fig. 4 where the inhibitory potency of the photosynthesis-inhibiting herbicide *metribuzin* has been measured with isolated chloroplasts. The example also illustrates the reproducibility of the data which allow to give the standard deviations together with the pI_{50}-value. The concentration required for 50% inhibition of a metabolic pathway should generally be in the μM range in order to contribute to the mechanism of action. However, exceptions to this rule are known (see e.g. Table 4).

A second requirement is the rapid appearance of the effect after herbicide application. This is illustrated in Fig. 5 where the drop in the concentration of soluble

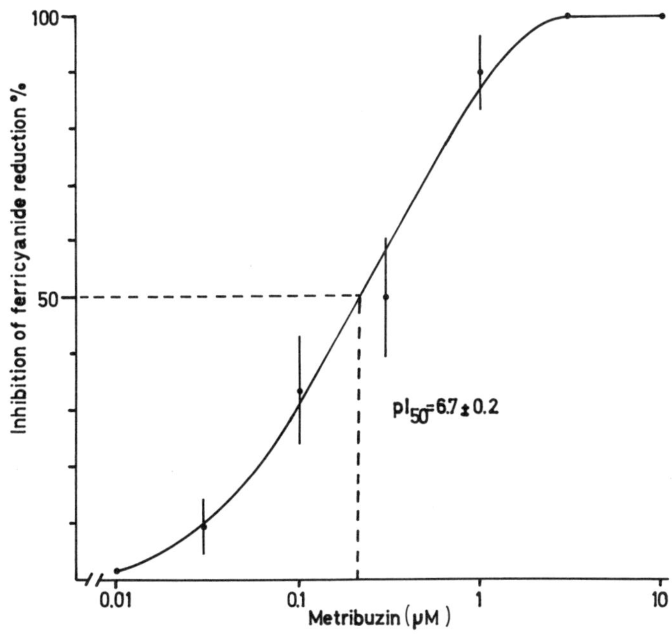

Fig. 4. Concentration dependence of the inhibition of electron flow in isolated chloroplasts by *metribuzin*. The pI_{50}-value (the negative decadic logarithm of the concentration needed for 50% inhibition) and the standard deviation are given

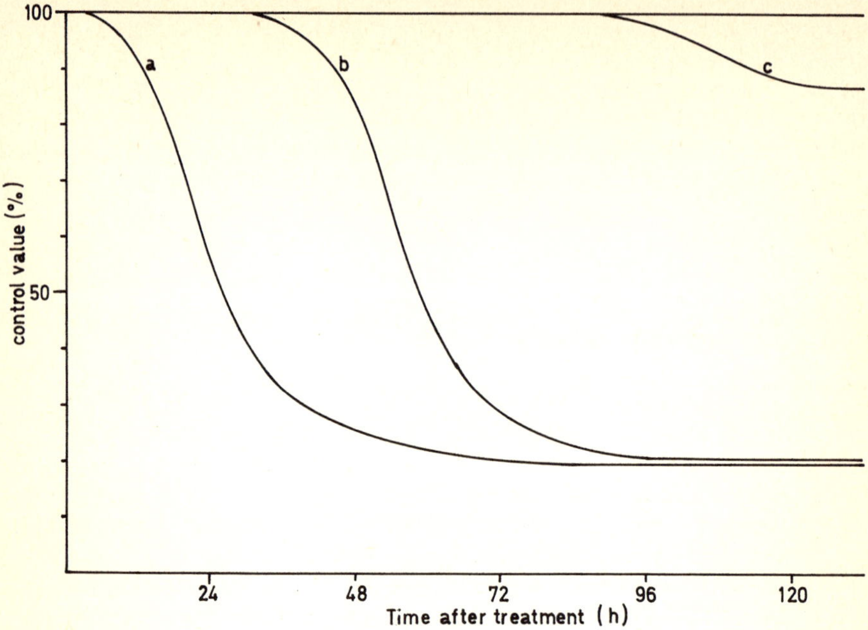

Fig. 5. Successive occurrence of direct and indirect metabolic changes after application of the photosynthesis inhibitor *methabenzthiazuron* to intact plants [1, 2]. *a* soluble reducing sugars, *b* malate, *c* chlorophyll $^{a}/_{b}$

reducing sugars (trace a) after *methabenzthiazuron* application occurs quite rapidly and can directly be related to photosynthesis inhibition. The other observed changes may then be considered as secondary responses (see p. 71). However, with some "slowly acting herbicides" the immediate effect on a sensitive metabolic step need not immediately show up in a measurable form (e.g. p. 187).

In Fig. 6 dose response curves of different herbicides in an in vitro test system are compared. When interpreting the data, both the inhibitory concentration range and the steepness of the curve should be taken into account. Poor concentration characteristics may imply unspecific interference. In Fig. 7 the reverse case is considered, i.e. the inhibition by one herbicide of different metabolic pathways to a different extent. At 1 μM only reaction (a) is inhibited, at 10 μM the inhibition of reaction (b) contributes to the overall measurable effects, and at 100 μM the situation may be very confusing because no clear cause-effect assignment is any longer possible. Such a situation is not uncommon, especially with "complex" and "supercomplex systems" (e.g. p. 148). However, since only the herbicide concentration in the tissue is relevant, the actual concentration applied externally might be misleading. One example is presented in Fig. 8, where the more specific test system (α-amylase synthesis) is less sensitive than seedling growth (trace a). Any conclusion drawn from the specific test system should therefore be treated with caution (e.g. concerning protein synthesis which participates in this system).

Stimulations from herbicidal influence do frequently occur, but they may originate from very different reasons. In Fig. 9 some examples are given. These include

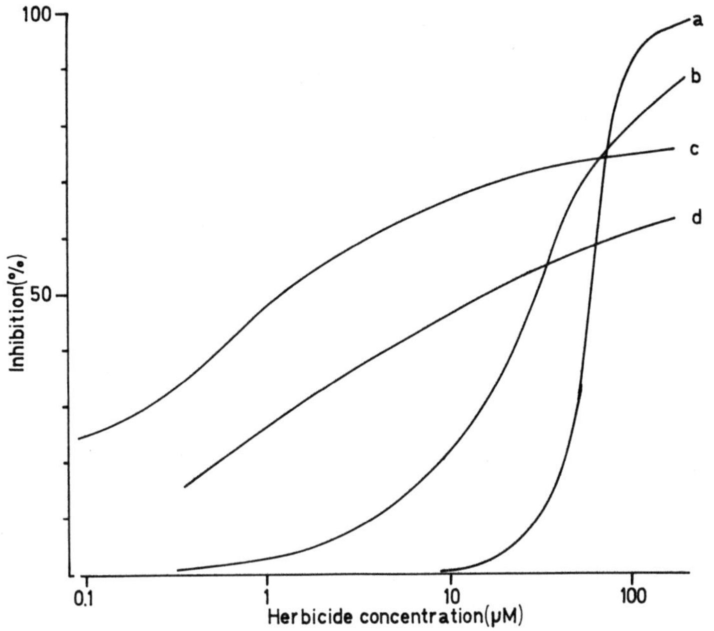

Fig. 6. Dose response curves of different herbicides for the inhibition of PMS-catalyzed cyclic photophosphorylation with isolated chloroplasts [3]. *a CIPC, b propanil, c diuron, d simeton*

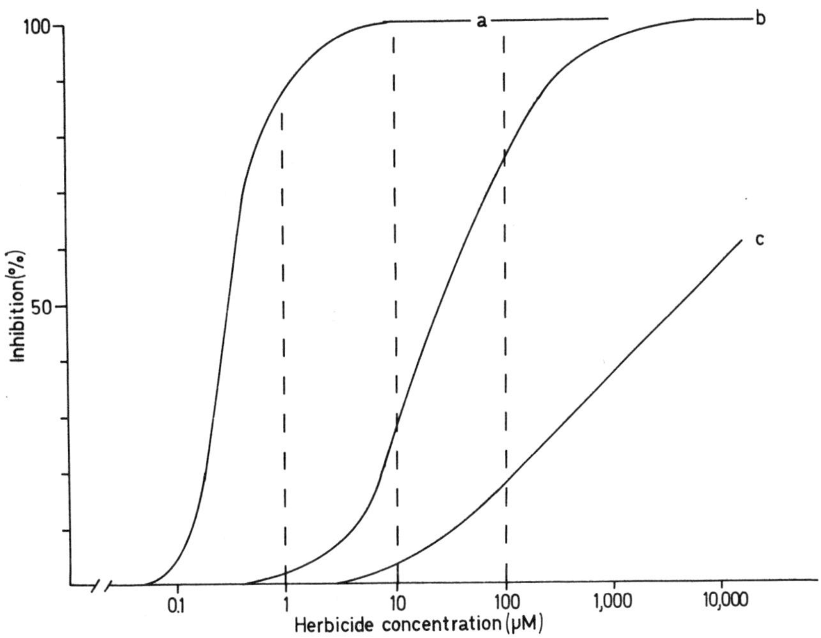

Fig. 7. Anticipated concentration-inhibition curves for a herbicide with 3 different independent sites of inhibitory action *a*, *b*, and *c*

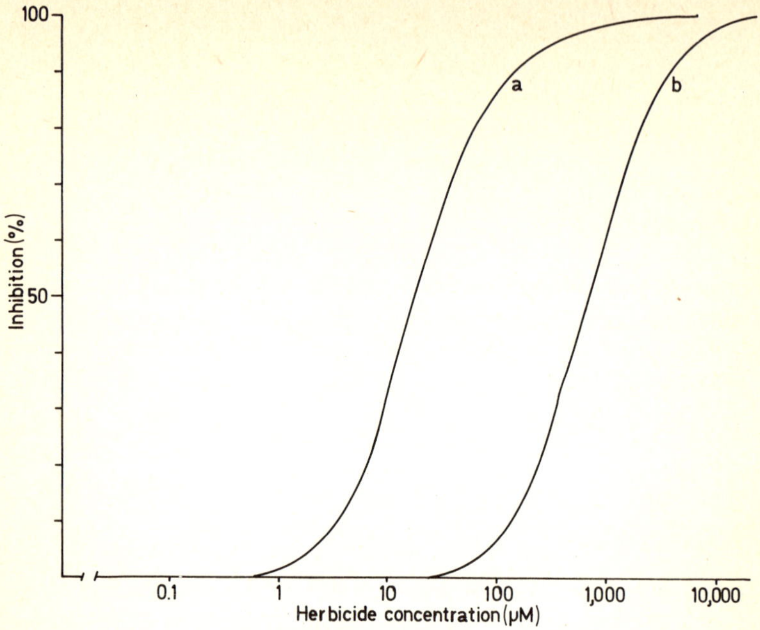

Fig. 8. Concentration-inhibition curves for *CDAA* inhibition of barley seedling growth (trace *a*) and of gibberellic acid-induced α-amylase synthesis in barley seeds (trace *b*) [4]. Reprinted in part with permission of the Journal of Agricultural and Food Chemistry. Copyright 1969 American Chemical Society

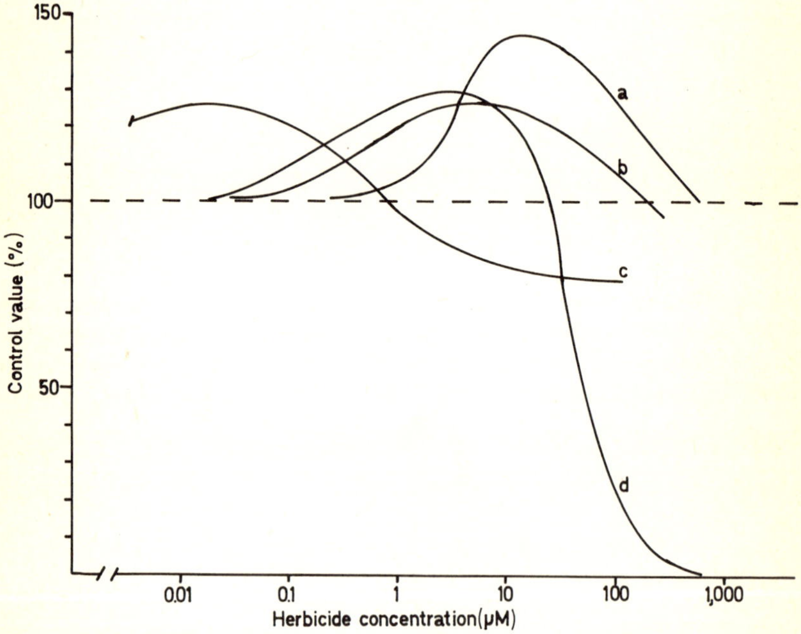

Fig. 9. Stimulation/inhibition concentration curves for different herbicidal effects. *a 2,4-D* induced nucleic acid increase [5], *b atrazine*-induced increase of ^{14}C-ATP incorporation into isolated chromatin [6], *c* growth response of bean embryonic axes to *atrazine* [7], *d* response of light-induced ATPase to *dinoseb* and other respiratory uncouplers [8, 9]. Reproduced with permission of the Weed Science Society of America

"direct" stimulations of the hormone type in the case of 2,4-D (trace a, possibly also trace b), regulatory stimulation responses after a more distant primary effect (trace c, see also p. 71), and a stimulation resulting from the elimination of a rate-limiting step (uncoupling effect, trace d, see also p. 115).

V. The Chain of Effects

When talking about the mechanism of action of herbicides, usually the inhibition of a sensitive pathway leading to the death of the weed plant is meant. As seen from a logistic viewpoint, this is an interdependency problem with three components: herbicide, measurable effects, and visible effects. Whereas modern laboratory

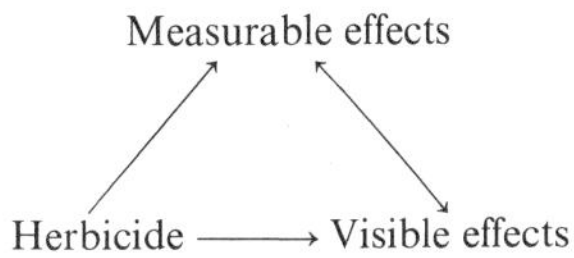

equipment and sophisticated physiological and biochemical methods allow a vast amount of specific data to be obtained, the information should finally be transformed into a logical sequence for the explanation of some very broadly described visible effects like bleaching, browning, wilting, yellowing, necrotic spots, desiccation or stunting, curling, swelling, twisting, and elongation of plant tissue. The measurable effects should be arranged in a sequence of cause-response relations with visible effects forming the end of this sequence. The scheme below schematically represents the chain of effects for inhibitors of

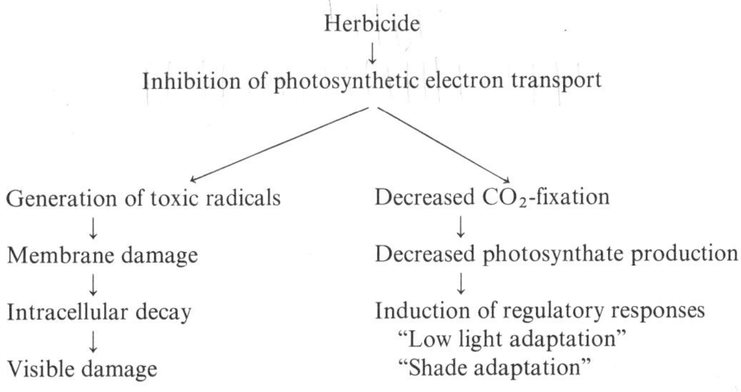

the photosystem II (cf. p. 24), e.g. *diuron* or *methabenzthiazuron*. This is a somewhat complicated example in that two independent visible effects originate from the same primary effect, i.e. the inhibition of photosynthetic electron flow. Whereas the visible damage in leaf tissue comes from unknown intermediates (toxic radicals or activated molecules, see p. 66), another chain of effects is initiated by the decreased photosynthate production, which itself is a consequence of the inhibited photosynthetic CO_2-fixation. I.e., the herbicidal damage is not caused by carbohydrate starvation, but the low carbohydrate supply causes regulatory responses. In the case of low herbicide concentrations which do not immediately kill the plant, there remains sufficient time for the plant to adapt itself to the decreased sugar supply. After 3–5 days there will be profound changes in nitrogen metabolism and chloroplast composition which are typical of shade-adapted plants. The symptom best visible is an increased chlorophyll concentration, the so-called "greening effect" (p. 73). These phenomena are also called "sublethal effects" or "physiological effects". In this case the chain may even further be completed by "ecological effects": it is well-established that the decreased sugar concentrations in the treated plants encourage so-called "low-sugar deseases" *(Alternaria, Fusarium, Helminthosporium = Drechslera)*, whereas at the same time "high-sugar deseases" *(Botrytis, Verticillium, Cercosporella)* show a reduced rate of infection. Additionally, the simultaneously increased amino acid concentration might be the reason for the preference by aphids of these plants, since aphids are known to select feeding plants for high amino acid levels.

When putting together the jigsaw of information pieces, the picture is still very incomplete. Part of the scene already emerges, but there are pieces for which we don't know the place, other pieces might not belong to the jigsaw, and of course many pieces are still missing. Many survey articles on herbicide chemistry and modes of action have been published [10–24].

VI. Mobility and Selectivity of Herbicides

Mobility and selectivity of herbicides will be considered in this treatise only as far as the available knowledge contributes significantly to our understanding of the mechanism of action. Compounds entering via the root diffuse through the outer cell wall and then find their barrier, the plasma membrane (Fig. 10). This basically lipohilic structure with hydrophilic inner and outer faces contains pores, which, however, are inhabited by proteins, the latter being possibly catalysts for active transport systems. Since these gates into the cell's interior are fairly selective in which molecules they allow to pass, most herbicide molecules will simply diffuse through the lipid bilayer. Considering the fact that most herbicides exhibit at least partially lipophilic structures, they dissolve into the lipid bilayer and reemenate on both sides, inside and outside. This means that the plant apparently is not able to exclude foreign molecules of a certain size which are not heavily hydrated. By contrast, inorganic ions surrounded by a heavy hydrate sheath like K^+, Na^+, Ca^{2+},

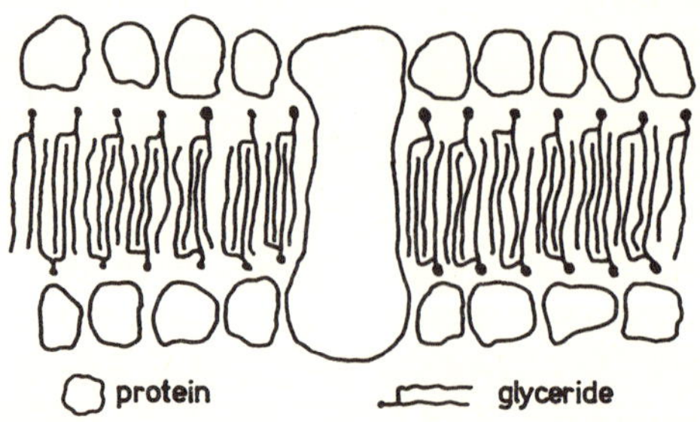

Fig. 10. Membrane model including a lipophilic interior, hydrophilic external proteinaceous faces, and a proteinaceous pore

Table 2. Examples for selective actions of herbicides

Herbicide	Selectivity by	Page
Metamitron	Rapid detoxification	41
Fluometuron		32
Propanil etc.		34
2,4-D	Low uptake and low mobility	171
Bifenox		94
2,4-D	Different physiological response	172
Paraquat		89
Chlorfenprop-methyl	Difference at the site of action	180
Atrazine	Mutation at the site of action	55
Metflurazone	Physical reasons	110
Metribuzin	Applicational reasons	43

or NO_3^-, SO_4^{2-}, HPO_3^{2-} etc. appear not to be able to pass through the lipid bilayer but are taken up by specific transport systems.

A diffusion of herbicides by passage through the cell wall system is stopped at the casparian strip, an ingenious construction which blocks all water flow through the cell wall system and into the central root cylinder. This structure is basically a waxy strip surrounding the cells of the endodermis layer. All molecules taken up by the root must therefore pass through the living cytoplasm and may be detoxified or deposited there.

Once a herbicide has entered the central cylinder and has been passed into the conductive tissue, is moves passively upwards in the xylem with the transpiration stream. In order to reach its site of action, e.g. the chloroplast of the leaf, the herbicide has again to penetrate through a number of membranes, and it may be detoxified during its way through the cytoplasm.

Detoxification by the tolerant plant is the most common basis for selectivity in the case of xylem-mobile herbicides (Table 2). Selectivity S for these compounds can be described by the function

$$S = \frac{\text{detoxification}}{\text{uptake} \times \text{transport}}.$$

As long as the capacity for detoxification keeps pace with the amount of incoming herbicide (uptake × transport), the treated plant proves to be tolerant towards this herbicide. Therefore, a low or high capacity for detoxification in itself is not sufficient for the determination of sensitivity or tolerance, but should always be seen together with the rates of uptake and transport. However, the detoxification capacity has by far the greatest influence on selective herbicidal action among these three parameters. A number of other possible reasons for selective action is listed in Table 2.

References for A

1. Fedtke, C.: Ber. Deutsch. Bot. Ges. *87*, 155 (1974)
2. Fedtke, C.: Pestic. Biochem. Physiol. *4*, 386 (1974)
3. Nishimura, K. et al.: Agr. Biol. Chem. *39*, 867 (1975)

4. Jaworski, E.G.: J. Agr. Food Chem. *17*, 165 (1969)
5. Chen, L.G., Switzer, C.M., Fletcher, R.A.: Weed Sci. *20*, 53 (1972)
6. Penner, D., Early, R.W.: Weed Sci. *20*, 367 (1972)
7. Bush, P.B., Ries, S.K.: Weed Sci. *22*, 227 (1974)
8. Alsop, W.R., Moreland, D.E.: Pestic. Biochem. Physiol. *5*, 163 (1975)
9. Moreland, D.E., Hussey, G.G., Farmer, F.S.: Pestic. Biochem. Physiol. *4*, 356 (1974)
10. Moreland, D.E.: Ann. Rev. Plant Physiol. *31*, 597 (1980)
11. Jäger, G.: Unkrautbekämpfungsmittel (Herbizide). In: Büchel, K.H. (ed.): Pflanzenschutz und Schädlingsbekämpfung, p. 155. Stuttgart: Thieme 1977
12. Kühle, E.: Herbizide. In: Ullmanns Enzyklopädie der technischen Chemie, p. 597. Weinheim: Verlag Chemie 1976
13. Draber, W., Fedtke, C.: Herbicide interaction with plant biochemical systems. In: Geissbühler, H. (ed.): Advances in Pesticide Science, part 3, p. 475. Oxford-New York: Pergamon 1979
14. Anderson, J.L., Thomson, W.W.: The effects of herbicides on the ultrastructure of plant cells. In: Gunther, F.A. (ed.): Resid. Rev., p. 167. New York-Heidelberg-Berlin: Springer 1973
15. Corbett, J.R.: The biochemical mode of action of pesticides, pp. 330. London-New York-San Francisco: Academic Press 1974
16. Frear, D.S. et al.: Agronomy *24*, 327 (1972)
17. Gräser, H.: Biochemie und Physiologie der Phytoeffektoren, pp. 446. Weinheim: Verlag Chemie 1977
18. Dodge, A.D.: Sci. Progress (Oxf.) *62*, 447 (1975)
19. Ashton, F.M., Crafts, A.S.: Mode of action of herbicides, pp. 504. New York-London-Sydney-Toronto: John Wiley and Sons 1973
20. Audus, L.J. (ed.): Herbicides-physiology, biochemistry, ecology, Vols. I, II. London-New York-San Francisco: Academic Press 1976[2]
21. Fryer, J.D., Makepeace, R.J.: Weed Control Handbook, Vols. I, II. Oxford-London-Edinburgh-Melbourne: Blackwell Scientific Publ. 1977[6]
22. Wegler, R., Eue, L.: Herbizide. In: Wegler, R. (ed.): Chemie der Pflanzenschutz- und Schädlingsbekämpfungsmittel, Vol. 2, p. 172. Berlin-Heidelberg-New York: Springer 1970
23. Wegler, R., Eue, L.: Herbizide. In: Wegler, R. (ed.): Chemie der Pflanzenschutz- und Schädlingsbekämpfungsmittel, Vol. 5. Berlin-Heidelberg-New York: Springer 1977
24. Pfister, K., Urbach, W.: Effects of biocides and artificially induced growth regulators: Physiological basis. In: Pirson, A., Zimmermann, M.H. (eds.): Encyclopedia of Plant Physiology, Vol. 12 D, Physiological Plant Ecology IV, Berlin-Heidelberg-New York: Springer, in press

B. Plant Metabolism – a Synopsis of Principles

I. Plant Growth as an Open System

Plant growth is basically a system with two opposite polarities (Fig. 11): the shoot growing against gravity (negative gravitropism) and/or into the direction of incoming light (phototropism) and the root growing in the direction of gravity pull (positive gravitropism). Lateral roots and lateral branches hold a balance between gravitropic, phototropic, and other directional responses. Directional growth is brought about by two opposite meristems delivering cells for the build-up of the plant. When the cells begin to fulfill their function in the developing tissue they differentiate from the isodiametric meristematic cell filled with cytoplasm and a big nucleus into the elongated mature differentiated cell containing a big central vacuole (Fig. 11). The nucleus is then no longer very prominent, but instead organelles have been differentiated from protoorganelles concerning the specific function(s) of the differentiated cell. By this type of growth the plant is basically immortable, and only size and hormonal correlations control the amount of new growth.

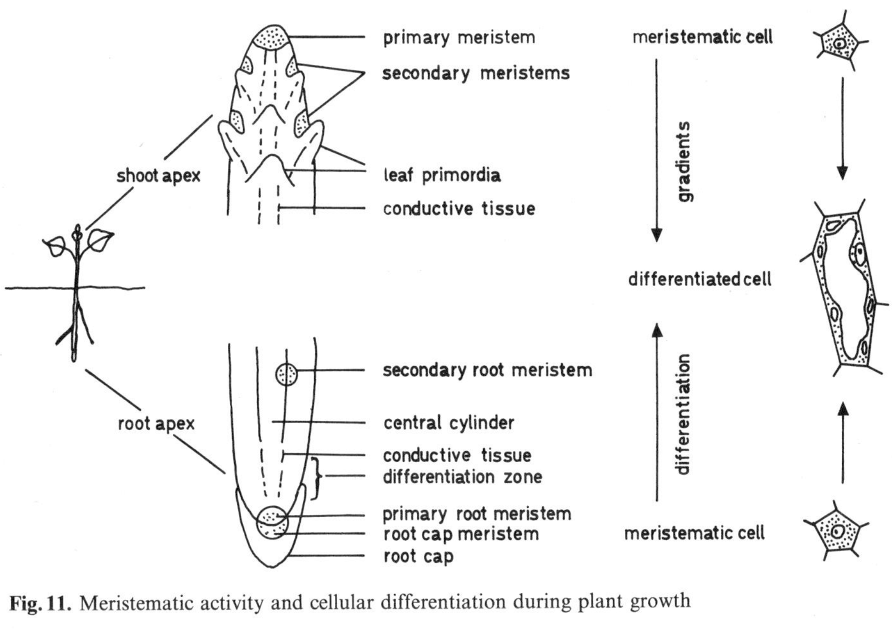

Fig. 11. Meristematic activity and cellular differentiation during plant growth

In the differentiation zone below the primary shoot meristem, secondary meristems develop at the sites of leaf primordia simply by remaining meristematic. In this process the entire shoot system of the plant holds a reservoir of meristematic regions which may stay even after leaf drop and allow a regrowth of the plant in the case of damage. However, completely differentiated cells may, under certain circumstances, dedifferentiate and again become meristematic. This occurs e.g. when potato tubers are injured and form a new bark; the differentiation of lateral root primordia from cells of the pericycle is another example. In callus culture systems differentiated cells of e.g. a carrot dedifferentiate into a meristematic conglomerate of cells; the callus then may be induced to form root and shoot meristems and to regenerate an intact plant by application of the appropriate hormone levels (auxin, cytokinin).

II. Cellular Organelles and Endomembrane Systems

Cell and subcell. All nucleate cells are divided into a number of compartments by endomembrane systems (Fig. 12). These compartments may be regarded as closed reaction vessels, but of course they also interact with other compartments and with the cytoplasm in between. Some of the more important compartments of the higher plant cell which will shortly be described here are the chloroplast, the peroxisome, the mitochondrion, the nucleus, and the vacuole.

The middle lamella is a product of cell division; it is formed in the division plane between the daughter cells (cf. p. 123). The cell wall then grows from both sides by intusception and apposition of wall material. The intercellular spaces remain open for aeration of the tissue.

Most of the space in the interior of the cell is inhabited by a large central vacuole. It has often been stated that the plant does not possess an excretory sys-

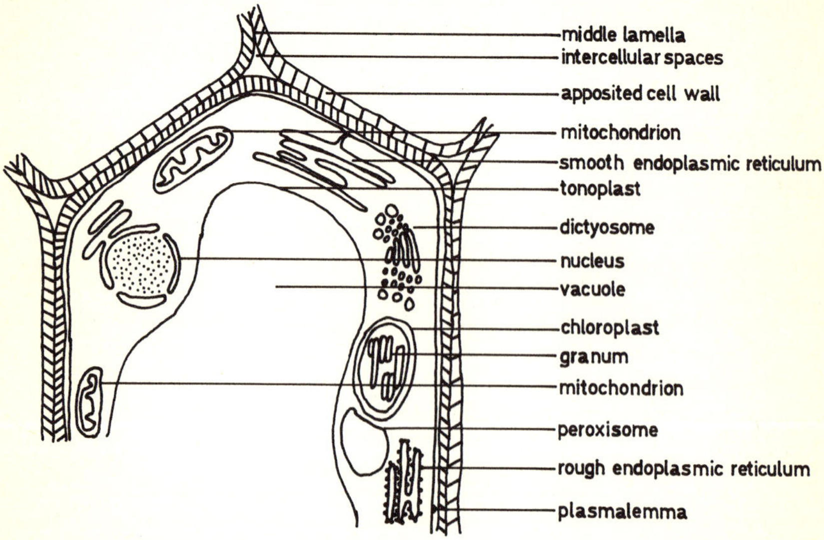

Fig. 12. Endomembrane systems in plant cells

tem, but this is not entirely true. The dictyosomes e.g. excrete the nectar of flowers. However, the plant cell apparently does not excrete metabolic waste products but instead deposits them in the vacuole.

Details on the present knowledge of plant physiology and biochemistry may be found in text books [1, 2].

The chloroplast. To carry out photosynthesis, the chloroplast is divided into subcompartments: grana lamellae, stroma lamellae, and stroma space. Whereas light-dependent reactions, i.e. the production of energy-rich compounds (ATP) and reduced coenzymes (NADPH) for the reduction of carbon dioxide to the level of carbohydrates, occur only in the grana and stroma lamellae, the fixation of CO_2 and the interconversions of carbohydrates including starch (dark reactions) proceed only in the stroma space. Grana and stroma lamellae have the form of flattened closed sacs called thylakoids and differ in their pigment content. Their interior spaces represent subcompartments (p. 114).

Peroxisomes. Peroxisomes are fairly large organelles surrounded by a single membrane. Peroxisomes and glyoxysomes are also called microbodies. The peroxisomal membrane often appears to be closely attached to the chloroplasts in electron micrographs. Since the peroxisomes contain the enzymes of the glycolic acid oxidation pathway, they are presumed to participate in the process called photorespiration (p. 20). They also contain catalase and are therefore able to destroy hydrogen peroxide (H_2O_2) which has been produced in other oxidative processes including glycolic acid oxidation.

The mitochondrion. In mitochondria as in chloroplasts different subcompartments are responsible for membrane-associated reaction sequences and for reactions occurring by the mediation of soluble enzymes. In the mitochondrion inner space compounds from the intermediate metabolism (Fig. 2) are broken down and CO_2 is formed. The reducing equivalents are transferred to the membrane-bound electron transport system for the generation of ATP (p. 114).

The nucleus. Not very many details are discernible in the interphase nucleus of the metabolically active cell. However, during the prophase of cellular division the chromosomes become visible after the nuclear membrane has been dissolved. A step very sensitive to some herbicides is the distribution of the chromosomes during cellular division (p. 123).

The endoplasmic reticulum. The endoplasmic reticulum (ER) occurs in two forms: smooth ER and rough ER. The surface of the rough ER is occupied by ribosomes and thereby turns into a place for protein synthesis. Phases of increased capacity for protein synthesis are characterized by an increased proportion of rough ER (p. 74). In addition, the membrane system of the ER appears to contain a number of bound enzyme systems and short electron-transport chains. During homogenization and separation on ultracentrifuge gradients the ER membranes are broken and turn up as the "microsomal fraction".

The vacuole. The vacuole is a multipurpose system serving at the same time as osmometer, storage compartment, water reservoir, waste container, and lysosome. Lysosomal activities of the vacuole are expressed by their content of hydrolytic enzymes with broad and varying activities (esterases, phosphatases, etc.). These hydrolases typically have rather acidic pH optima (around 5) which coincide with the pH in the vacuole. However, they may also be present as inactive precursor enzymes. During herbicidal action, rupture of the tonoplast and resulting cellular autolysis occur in some cases (p. 182).

References for B

1. Straßburger, E. et al.: Lehrbuch der Botanik für Hochschulen. Revised by Denffer, D. et al., pp. 1078. Stuttgart-New York: Gustav Fischer 1978[31]
2. Bonner, J., Varner J.E.: Plant biochemistry, pp. 925. New York-London: Academic Press 1976[3]

C. Photosynthesis

I. Physiology and Biochemistry of Photosynthesis

Photosynthesis depends on light. Looking deeper into this obvious dependence opens some important aspects of the overall photosynthetic process. In Fig. 13 the efficiency of photosynthetic CO_2-fixation in different regions of the visible spectrum of sunlight is compared with the absorption spectra of intact cells and of pure chlorophyll a and b. It is obvious that not only chlorophyll but also the absorption by other pigments contributes to the overall CO_2-fixation. In the chloroplast grana and stroma lamellae chlorophylls a and b are integrated with "accessory pigments" into the pigment systems I and II (p. 22). These accessory pigments include the yellow, orange, and brownish pigments β-carotene, lutein, violaxanthin, and neoxanthin.

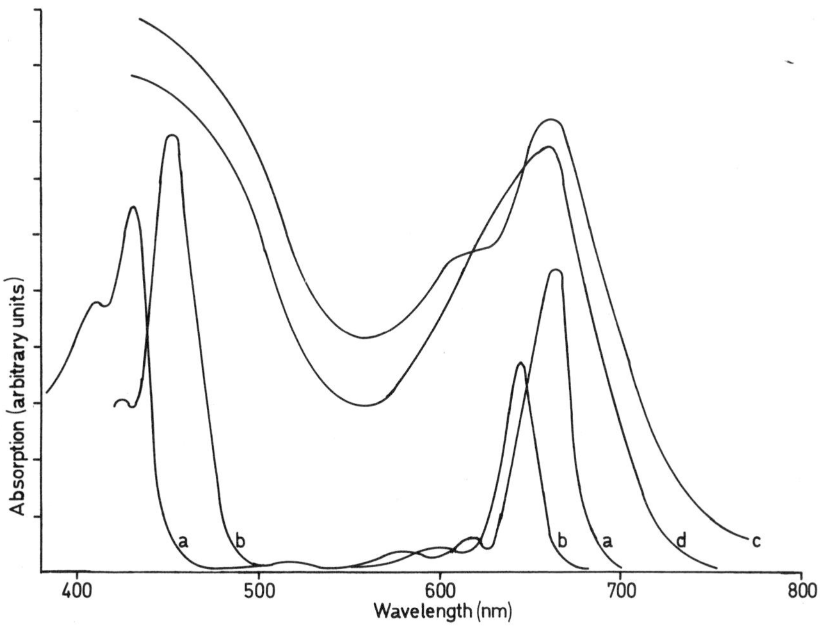

Fig. 13. Comparison of the absorption curves of chlorophylls *a* and *b* and intact *Chlorella* cells with the photosynthetic action spectrum of these cells. *a* chlorophyll a, *b* chlorophyll b, *c* whole cells, *d* action spectrum

Photosynthetic accessory pigments fulfill two equally important functions in the life and growth of photosynthetic organisms: (a) they extend the wavelengths of light benefited by photosynthesis from those absorbed by chlorophyll to their own absorption bands, and (b) they protect chlorophyll from lethal photooxidations (p. 63). The transfer of light energy absorbed by accessory pigments onto chlorophyll may be by charge transfer, fluorescence, or otherwise.

Light quality is only one important factor in photosynthesis. Other important external factors on which photosynthesis depends include light intensity, CO_2-concentration in the air, water supply, and temperature. However, only a few of these factors are relevant in studies on the mechanism of action of herbicides. One of these is light intensity which will be dealt with in Chap. C.VIII. Other factors, including those that are used to differentiate between C-3 and C-4 plants, e.g. photorespiration, are compared in Table 3. Photorespiration has occasionally been considered in studies on the mechanism of action of *s*-triazines (p. 35).

C-3 and C-4 plants have received their names because $^{14}CO_2$-fixation experiments have shown primary incorporation of radiolabel into either C-3 compounds (3-phosphoglyceric acid) or C-4 compounds (malate, aspartate). It appears, that C-4 metabolism represents a special adaptation by a number of species growing under warm and dry conditions; typical representatives are tropical grasses such as maize and sugar cane, but other plant families are also represented (e.g. *Chenopodiaceae*). The CO_2-fixation into C-4 acids appears to be a CO_2-concentrating mechanism, because the responsible carboxylating enzyme, phosphoenolpyruvate carboxylase (a), fixes CO_2 efficiently at much lower concentrations than the photosynthetic carboxylating enzyme, ribulosediphosphate carboxylase (b):

(a)

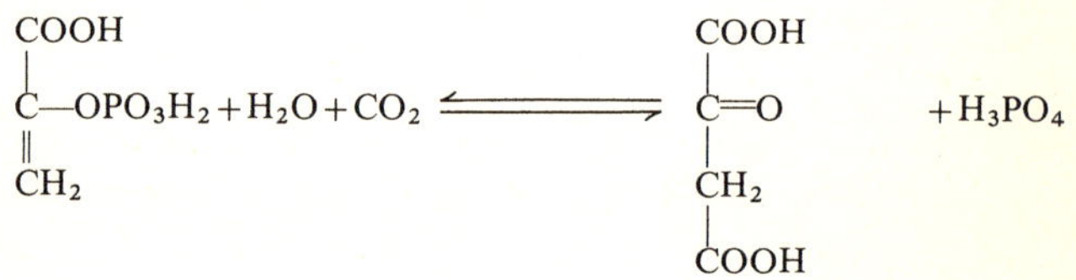

phosphoenol pyruvate oxaloacetate

(b)

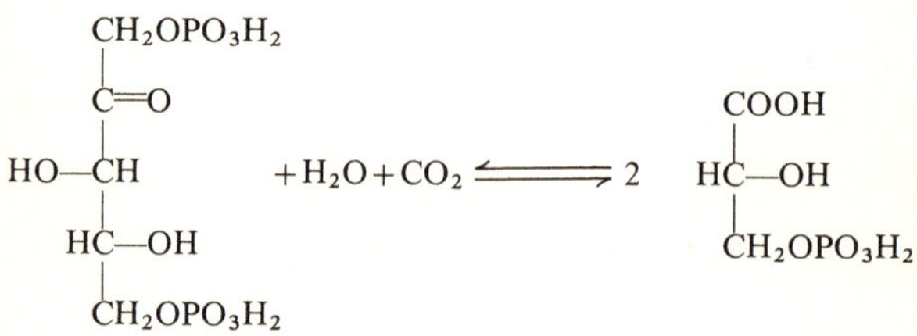

ribulose-1,5-diphosphate 3-phosphoglycerate

A special leaf anatomy in C-4 plants aids in the subsequent concentration of the carbon originally fixed in C-4 compounds (oxaloacetate, malate, aspartate) into the bundle sheath cells. There CO_2 is liberated and again refixed by the conventional C-3 pathway. These metabolic features, together with a very low level of

Table 3. Comparison of the characteristics of "C-3 plants" and "C-4 plants"

Parameter	C-3 plants	C-4 plants
Water requirement	High	Low
Temperature optimum	Low	High
Primary CO_2-fixation products	Sugars	Organic acids
Internal CO_2-enrichment	–	+
CO_2 compensation point	High	Low
Special leaf anatomy	–	+
Photorespiration	+	–

photorespiration, i.e. the liberation as CO_2 of part of the carbon originally fixed, lead to the low CO_2 compensation point generally found in C-4 plants. Though the C-4 type metabolism is of great selective advantage to many plants in their native habitat, C-3 metabolism is superior in cooler climates. Specifically, C-4 metabolism has a greater energy demand – in the form of ATP consumed for the CO_2-concentrating mechanism – which might be a disadvantage under temperate and humid conditions where C-3 plants dominate.

II. Flow of Electrons in Photosynthesis

The primary event in photosynthesis is the excitation of a pigment molecule after the absorption of a quantum of light energy. After this energy has travelled through the pigment complex to the "chlorophyll-trap" (P_{700} or P_{682} in Fig. 14), it may eventually be captured in an energized electron, leaving a cationic chlorophyll$^+$ (Fig. 26). The electron is replaced from water by an enzyme complex containing manganese, and the splitting of water leads to the liberation of molecular oxygen.

The energized electron derived from chlorophyll P_{682} may now reduce other compounds, depending on the redox potential. The native electron acceptor for electrons from photosystem II (PS II) is the quencher Q, a compound whose molecular identity is not yet clear. However, its name is derived from the observation that its function is related to fluorescence: excess excitation energy not used by the electron transport chain is given off as fluorescence or, conversely, efficient electron transfer to plastoquinone (PQ) quenches the fluorescence of Q. The fluorescence becomes greatest when the electron transport is blocked by a herbicide, e.g. *diuron* or *metribuzin* (p. 29 and Fig. 20). The sequence of participants in the photosynthetic electron transport chain as presented in Fig. 14 is arranged according to their redox potential. The redox jump generated by the absorption of light energy is at least one volt; two pigment systems are working in sequence in the generation of the redox potential necessary for the reduction of ferredoxin (Fd), a low molecular weight protein mediating electron transport at redox potentials around -300 to -450 mV. The exact redox potential and the exact molecular identity of X, the native electron acceptor for photosystem I, are not known. Electrons ac-

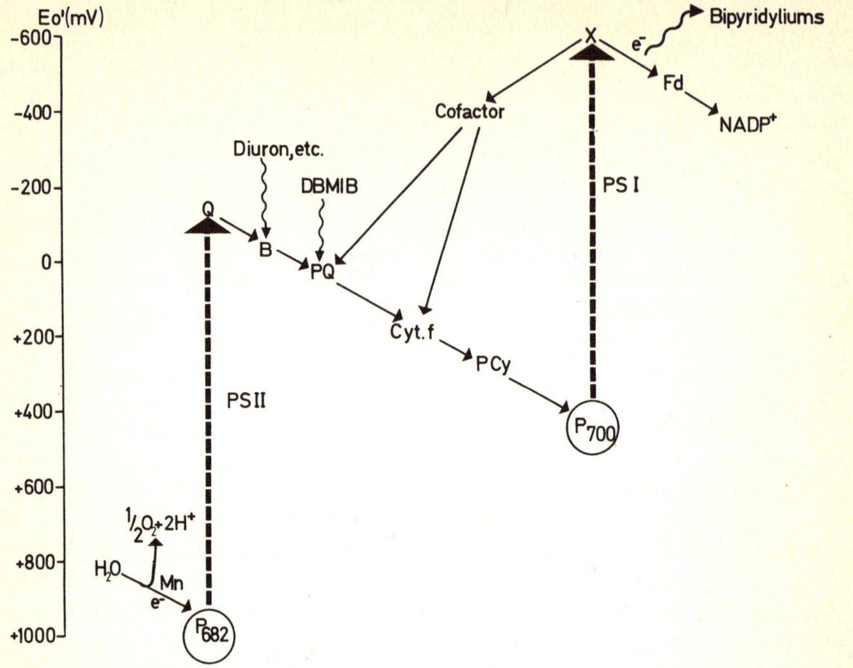

Fig. 14. The photosynthetic electron transport system, roughly ordered by the standard redox potentials (E_0) of the individual redox carriers [1, 2]. Inhibition/interference sites for different inhibitors of photosynthesis are indicated by waved arrows

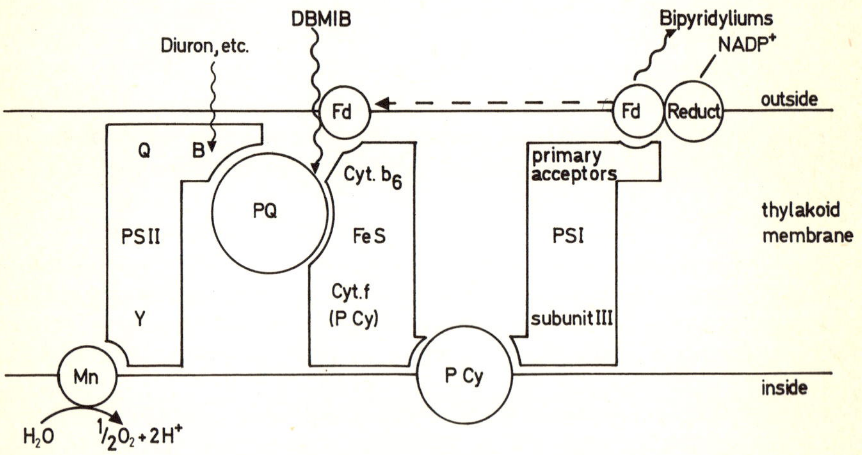

Fig. 15. Components of the photosynthetic electron transport system as they catalyze the vectorial electron flow across the membrane (cf. Fig. 14)

cepted by X – besides being transferred to ferredoxin – may also be transferred to the bipyridylium herbicides (p. 86) or may recycle by the mediation of endogenous or exogenous cofactors.

Whereas in Fig. 14 the photosynthetic electron flow is presented on a redox scale, Fig. 15 shows the same system as located in the thylakoid membrane. Oxygen is liberated in the thylakoid interior together with protons (H^+). A vectorial electron flow occurs across the membrane to the outside and again to the thylakoid inside via the plastoquinone shuttle. Plastoquinone in the reduced form transports two electrons and two protons whereas the subsequent participants in the electron transport chain, cytochrome f (Cyt f) and plastocyanine (PCy), only accept electrons. As a consequence, the thylakoid interior becomes acidic (the pH drops from ca. 7.5 to ca. 4), a process which is related to the conservation of energy in the form of ATP (p. 114). The final reduction of ferredoxin, of course, occurs on the stroma side of the thylakoid, and also the herbicides (*diuron*-like compounds and *bipyridyliums*) act on the outside.

The cyclic electron flow indicated in Figs. 14 and 15 may conserve energy in the form of ATP when it is coupled to a simultaneous proton transport into the thylakoid interior.

III. Inhibition of Photosynthetic Electron Flow

1. Action on the Isolated Chloroplast

The herbicide *diuron* is an appropriate example for the "career" of a metabolic inhibitor. Selected from an ordinary herbicide screening at Dupont in 1955, it was introduced as a herbicide for weed control in cotton. Soon its inhibitory potency in photosynthesis was detected [19, 20] along with other phenylurea herbicides with related structures. In effect, *diuron* played the key role in further research on photosynthetic electron transport and became familiar to plant biochemists and physiologists as DCMU (*di*chlorophenyldi*m*ethyl*u*rea). Since then, very few studies on the physiology or biochemistry of photosynthesis have avoided further use of DCMU.

The inhibitory potency of a herbicide in photosynthetic electron transport is expressed by the pI_{50}-value (Fig. 4). Since the exact pI_{50}-value is a function of the

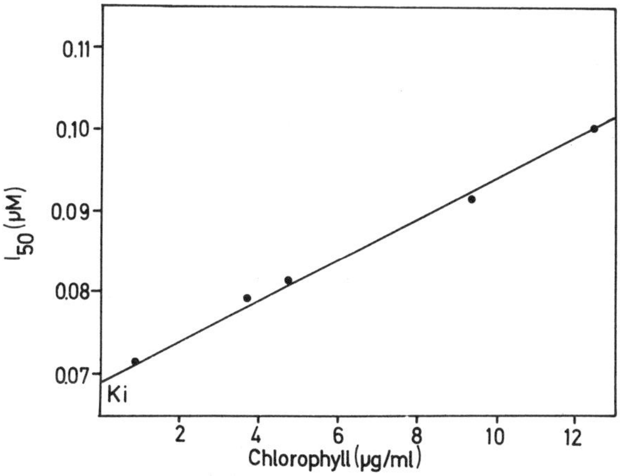

Fig. 16. Dependence of the I_{50}-value (µM concentration required for 50% inhibition of chloroplast electron transport) on the chlorophyll concentration in the test solution [21]

chloroplast concentration (Fig. 16) it can not always be compared from different test systems. Furthermore, the use of different electron donors or acceptors might include only part of the electron transport chain in the assay (Fig. 14). Testing only limited sections of the electron transport chain also allows the exact site of inhibitory action to be located [1]. For *diuron* this site is the reducing side of the "quencher Q" [1, 22–24], more precisely the reoxidation of the reduced Q (primary acceptor for photosystem II) by B (hypothetical specific plastoquinone between Q and PQ, Figs. 14, 15). Inhibition may occur by redox poising, i.e. by lowering the midpoint potential of B [25]. At elevated concentrations (0.1 mM), cyclic photophosphorylation may become uncoupled [23]. At lower concentrations, however, cyclic photophosphorylation is stimulated because the competition between open and cyclic electron transport is eliminated (Fig. 21). *Diuron* may even be used to determine the amount of cyclic photophosphorylation in vivo in algae [26]. From these studies it became clear that cyclic photophosphorylation is saturated at lower light intensities than photosynthesis and that it contributes from 10 to 50% of the ATP needed for the photosynthetic fixation of CO_2.

Besides the "*diuron* site", there also exists a "DBMIB-site" (Figs. 14, 15, Table 15) [27]. The DBMIB-site concerns the other "side" of the plastoquinone pool which extends through the membrane. Only slight structural changes may convert a "*diuron* type inhibitor" into a "DBMIB-type inhibitor", the latter e.g. including diphenyl ethers [28, 29] and dinitroaniline compounds [29] at elevated concentrations.

2. Herbicides Inhibiting at the "Diuron Site"

a. General Aspects

Herbicides inhibiting photosynthetic electron transport at the "*diuron* site" near photosystem II constitute about half of the herbicidal structures currently known. They may conveniently be grouped into classes of similar structures known as urea derivatives (Table 4), anilides (Table 5), *s*-triazines (Table 6), *as*-triazinones (Table 8), uraciles (Table 9), biscarbamates (Table 10), pyridazinones (Table 11), hydroxybenzonitriles (Table 12), nitrophenols (Table 13), and benzimidazoles (Table 14). Structures not included in these main classes are compiled in Table 15.

In these tables the structures are listed together with the water solubility and with the pI_{50}-values as obtained in isolated chloroplasts. References to the pI_{50}-values are given in the text when the compound concerned is first mentioned. A rough correlation exists between water solubility, pI_{50}-value, and the amount of herbicide to be applied in the field (Fig. 17). This correlation indicates (a) the mobility in the plant required by the herbicide in order to reach its site of action in the chloroplast and (b) the limited value of the pI_{50}-value to predict good herbicidal action. More precisely, three groups may cautiously be discerned: compounds with very good water-solubility and high pI_{50}-values requiring only low application rates in the field (e.g. *metribuzin*), compounds with low water-solubility but high pI_{50}-values necessitating normal application rates (e.g. *methabenzthiazuron*),

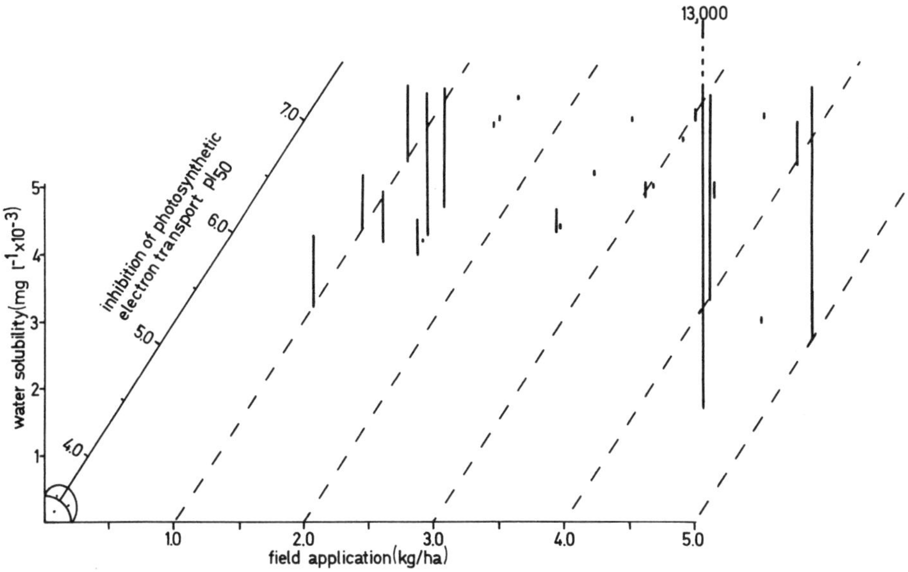

Fig. 17. Three-dimensional graph relating water solubility (lengths of the bars) with field application rates and pI_{50}-values for 29 herbicidal inhibitors of photosystem II

and compounds with high water-solubility but low pI_{50}-values were fairly high application rates are required in the field (e.g. *fenuron, metamitron, isocarbamid*). Individual herbicides among these "photosynthesis inhibitors" deviating considerably from this interdependence between water solubility, pI_{50}-value, and application rate should possibly be reexamined for other influences.

b. Urea Herbicides

Among urea herbicides (Table 4) the following compounds have received special attention:

Benzthiazuron [30] is a compound for weed control in sugar beet fields. Yet photosynthesis in these plants is as sensitive as in the susceptible weed [31]. However, the uptake of the herbicide by beet plants is comparatively low; this observation suggests reduced *benzthiazuron* uptake as a cause for selectivity.

Buthidazole [32–36] inhibits photosynthesis but not respiration. The tolerance has been related to differences in mobility.

Chlorbromuron [37] has been examined for its absorption by isolated cells or protoplasts from immature tomato fruit, but none was found [38, 39].

Chloroxuron [37, 40] was tested for a possible influence on the seed oils in soybean, but none was found [41]. However, *chloroxuron* increases the protein content and protein yield of rice plants when applied at the boot stage [42]. The significance of these stimulatory effects will be discussed in detail (p. 71).

Chlortoluron [37], which is applied for weed control in wheat and barley, is a selective herbicide because of its rapid detoxification in the resistant crop (wheat) [43].

Table 4. Structures, water-solubilities, and pI_{50}-values of urea herbicides ($-\log_{10}$ of the molar conc. inhibiting electron transport in isolated chloroplasts by 50%)

Common name	Structure	Water sol. ($mg\ l^{-1}$)	pI_{50}
Benzthiazuron	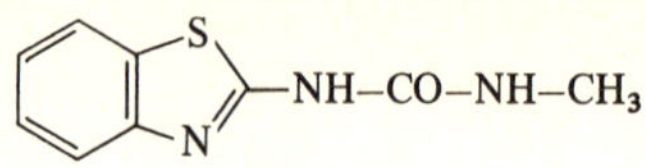	12	6.4
Buthidazole	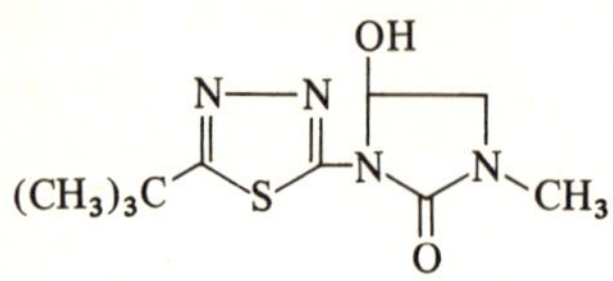	3,400	–
Buthiuron	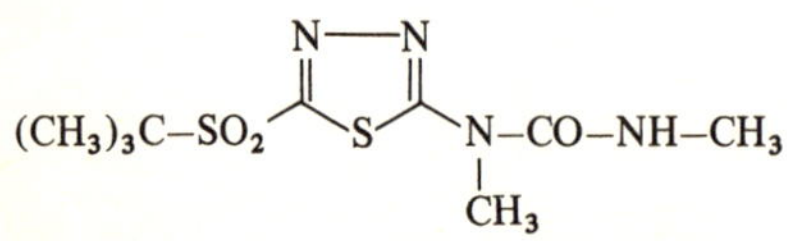	30	–
Buturon	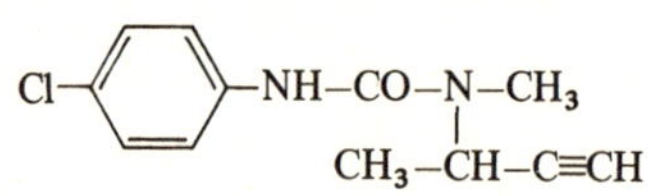	30	–
Chlor-bromuron	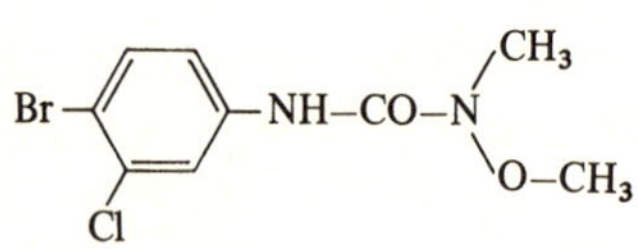	50	7.2
Chloroxuron	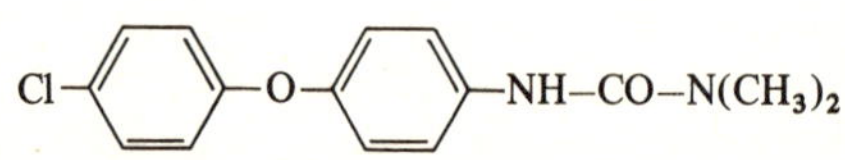	3.7	6.8–7.3
Chlortoluron	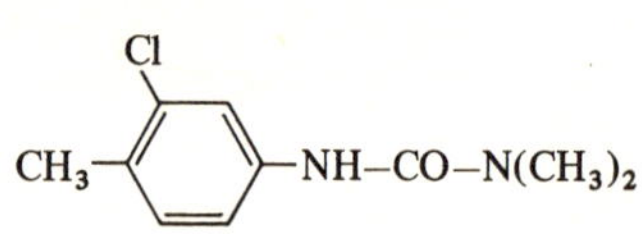	10	7.0
Cisanilide	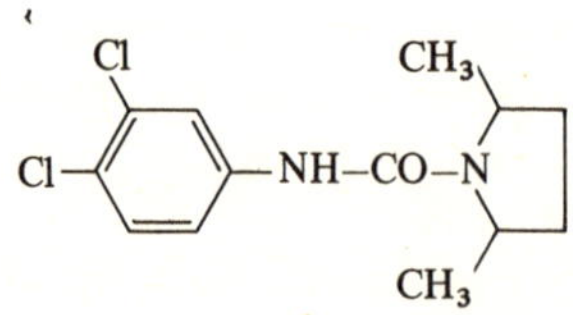	–	6.8
Cycluron	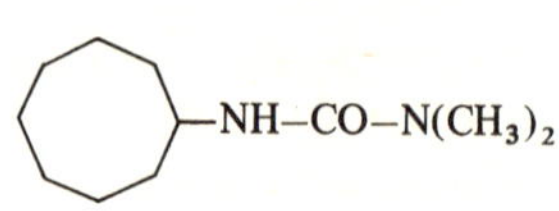	1,100	5.3
Dimefuron	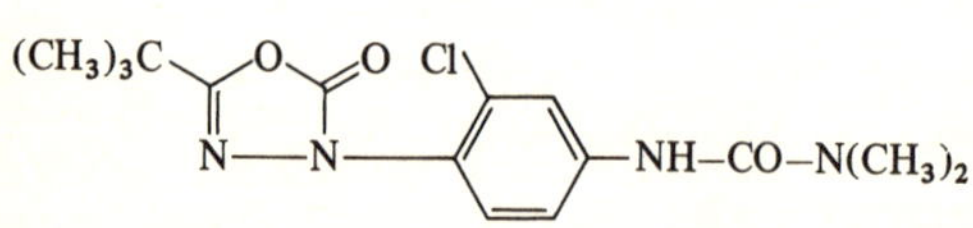	–	6.6

Table 4 (continued)

Common name	Structure	Water sol. (mg l^{-1})	pI_{50}
Diuron	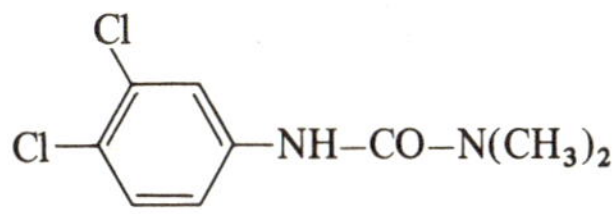	42	6.7–7.5
Ethidimuron	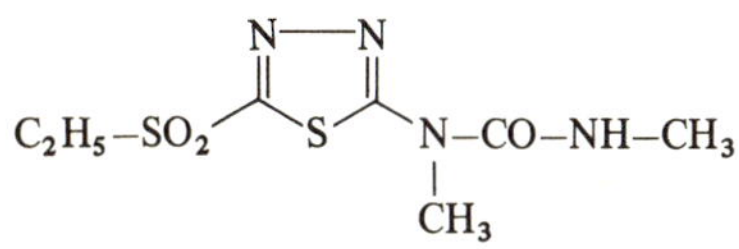	3,000	6.8
Fenuron		3,850	4.6–5.5
Fluometuron	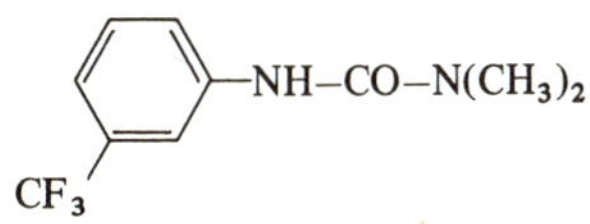	80	6.0
Isocarbamid	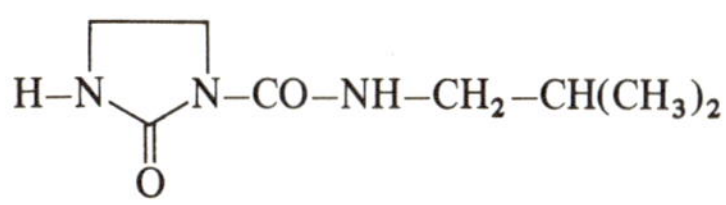	13,000	4.4
Isoproturon	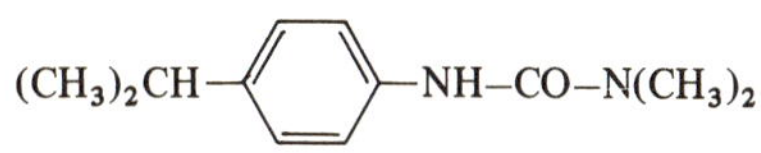	170	6.8
Linuron	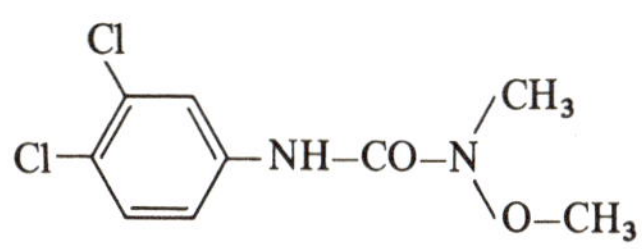	75	6.7–7.0
Methabenz-thiazuron	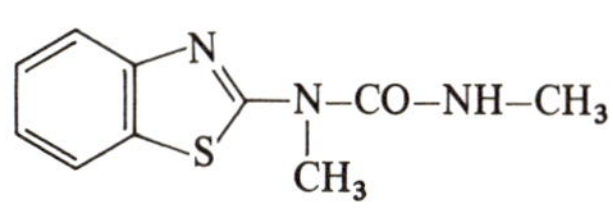	59	6.5
Methazole	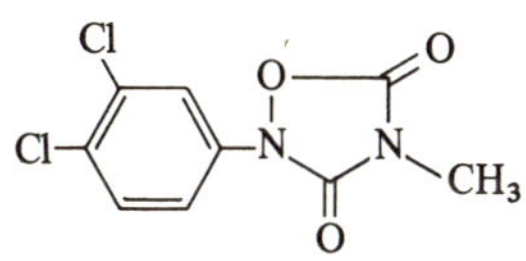	1.5	
Metobromuron	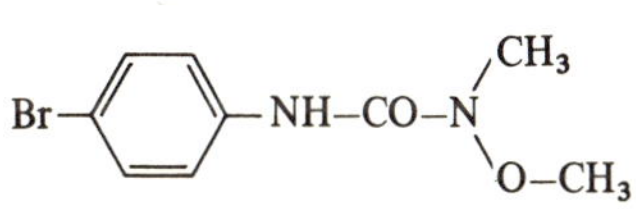	330	6.0

Table 4 (continued)

Common name	Structure	Water sol. (mg l^{-1})	pI_{50}
Metoxuron	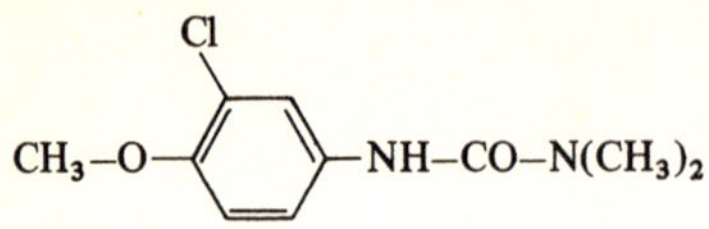	678	6.6
Methurin	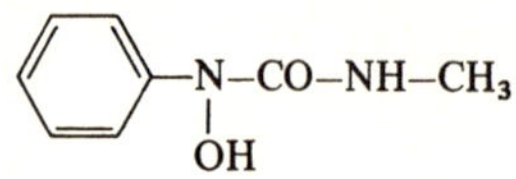	–	3.1
Monolinuron	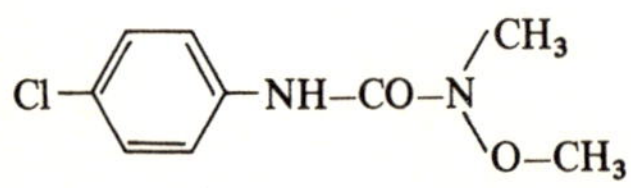	580	5.6–6.1
Monuron	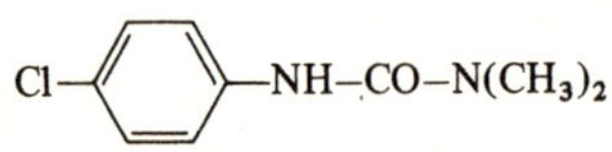	230	5.6–6.8
Neburon	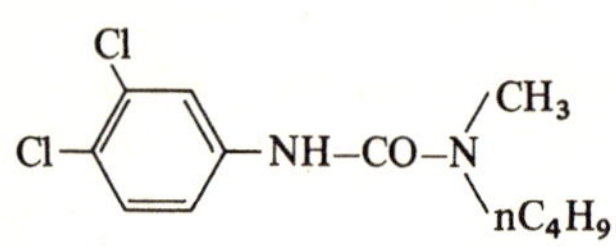	5	6.7–6.9
Noruron =norea	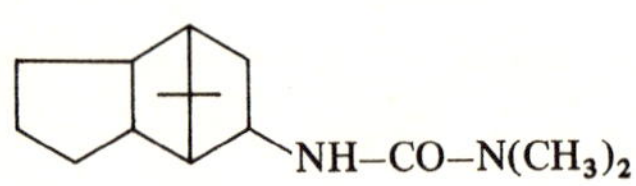	150	5.8
Siduron	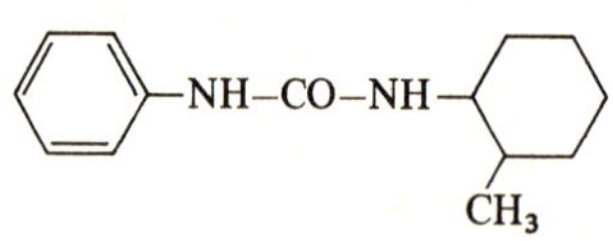	18	5.2
Tebuthiuron	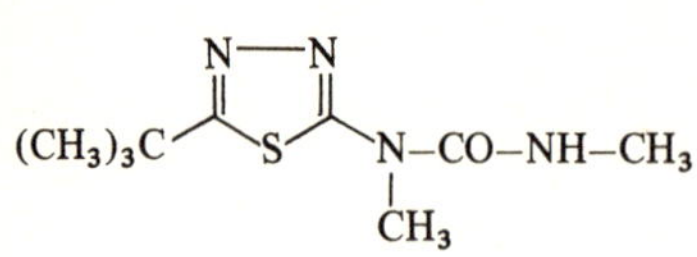	2.5	6.9
Thiochlor-methyl	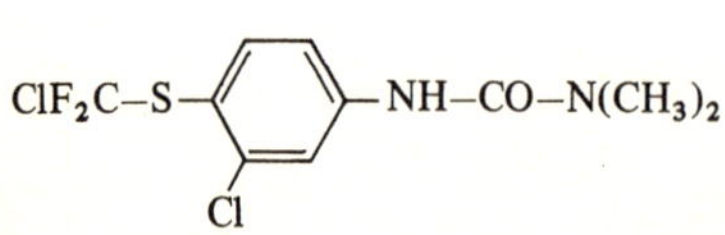	10	7.3

Cisanilide [44] is active as a herbicide only in the cis conformation. Here again, as with most of the xylem-mobile photosynthesis-inhibiting herbicides, rapid detoxification in the resistant plant supplies an explanation for the selective action. The herbicide antidote 1,8-naphthalic anhydride at the same time reduces the damage to the crop plant from *cisanilide* and increases its breakdown in the treated plant, due to increased activities of the relevant enzymes.

Cycluron [40].

Dimefuron [37] tolerance in field bean and pea has been investigated and found to be due to depth protection and additional root retention, respectively [45].

Diuron, often referred to as DCMU [19, 40, 44, 46, 47], is the standard compound used in plant physiological and biochemical laboratories around the world when inhibition of photosynthesis is required. The exact site of inhibitory action in the chloroplast electron-transport system has been thoroughly studied, and there is general agreement that this is the reducing side of photosystem II, more precisely the reoxidation of the reduced quencher Q (p. 21 and Figs. 14, 15). The available evidence on the exact inhibition site has been recently summarized by Izawa [24] and Trebst [1]. However, the exact inhibition site in filamentous cyanobacteria (blue-green algae) may differ from that in other photosynthetic organisms [48, 49]. All the herbicides discussed in this chapter (Tables 4–14) appear to inhibit the same step in photosynthetic electron transport and, moreover, obviously bind to the same binding site [2, 21]. The evidence available on the structure and localisation of the binding site is dealt with elsewhere (p. 51). There, too, resistance phenomena due to changes by mutation at the binding site are discussed.

Whereas the inhibition by *diuron* of photosynthetic electron transport at the reducing side of photosystem II is very specific, elevated concentrations cause inhibition also at other sites. Especially, the uncoupling of cyclic photophosphorylation by 1–10 μM *diuron* [47] (Fig. 18) deserves attention. Generally, the photosynthetic electron-transport system appears to be ten times more sensitive towards photosystem II herbicides than cyclic photophosphorylation. The reverse is true for the respiratory uncouplers (p. 115, Fig. 37).

At about 1 μM *diuron*, cyclic photophosphorylation processes might even be enhanced because the open electron transport from H_2O to NADP no longer competes with the cyclic electron transport (Fig. 21). For the understanding of this effect, it is relevant to envisage that the cyclic and noncyclic photosynthetic electron transport systems in part use the same electron transport chain (Fig. 14).

Structure activity correlations for *diuron* analogues have been studied [40, 50]. The involvement of steric parameters in the inhibition of photosynthetic electron transport became evident from experiments with N-(α-methylbenzyl)-N′-(3,4-dichlorophenyl)urea (MBPU): only the S-isomer revealed an inhibitory effect, whereas both the S- and the R-isomer inhibited all of the photophosphorylation reactions studied. On the urea moiety of the molecule, at least one free -NH- is required for inhibitory potency whereas the inhibition is lost with all the methyl groups [40]. However, the requirement for at least one free -NH- is not true for other classes of photosystem II inhibitors, not even for all urea herbicides [30].

As with other photosynthesis-inhibiting herbicides, with *diuron* too it is possible to increase the amount of nitrogenous compounds in the plant [51]. Oats, soybeans [52], and cotton respond to applications of sublethal *diuron* concentrations

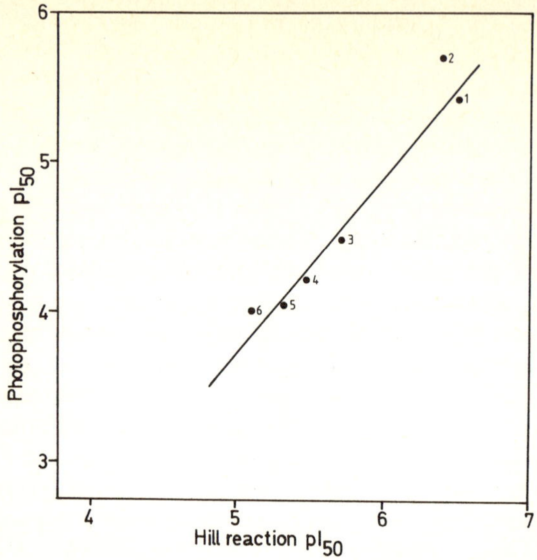

Fig. 18. Comparison of pI_{50}-values of electron transport inhibition with those of uncoupling of cyclic photophosphorylation for different photosystem II inhibitors [47]. *1 diuron, 2 simetryne, 3 propanil, 4 monolinuron, 5 swep, 6 simeton*

by increased contents of soluble protein, nitrate, and amide-nitrogen. (For a discussion of these effects see p. 70.)

As mentioned before, *diuron* has been employed as a standard photosynthesis inhibitor in many physiological experiments. It has e.g. been demonstrated, that chloride uptake by *Vallisneria spiralis* L. is only inhibited by 60% whereas CO_2-fixation is inhibited by 100% in the presence of 1 µM *diuron* [53]. From this result it has been assumed that the energy supply for active ion-uptake may come, at least in part, from cyclic photophosphorylation. Flowering in the long day plant *Lemna perpusilla* was enhanced by 1 µM *diuron*, suggesting that flowering in these plants is controlled by the rate of photosynthesis [54]. Since an interdependence of membrane potential, phytochrome red/dark red reversibility and photosynthesis rate has been demonstrated for the related *Lemna paucicostrata* [55], again by the application of 1 µM *diuron*, a flowering dependence on the phytochrome system might show up here. However, flowering in the long-day plant *Anagallis arversis* was reduced by $\geqq 1$ µM *diuron* [56]. Interestingly, sucrose completely antagonized the effect of *diuron*.

Another influence seen in intact plants is the decreased rate of transpiration [57, 58]. This is not only true for *diuron* but also for other inhibitors of photosynthesis [59, 60]. In general, the inhibition of transpiration will be less than the inhibition of photosynthesis. The cause for the effect on transpiration is presumed to be a stomatal closure due to a rise of the internal CO_2-concentration which, in turn, is induced by the inhibition of photosynthesis.

Many experiments have been performed with algae. Suspensions of unicellular algae ("microalgae") are the favorite object for many physiological experiments because of their synchronized growth, steady availability and rather low complexity. Galactolipid synthesis has been studied in *Chlorella* [61] and *Euglena* [62] and has been found to be inhibited by *diuron* in the light but not in the dark in *Chlorella*. Galactolipid synthesis for the chloroplast should therefore occur in the cytoplasm and depend on photosynthesis only indirectly. However, motility in unicellular algae may be shown by similar experiments to be either dependent on (*Eudorina elegans* [63]) or independent of (*Cryptomonas* sp., *Euglena gracilis*, *Chlamydomonas reinhardii* [64]) photosynthesis. Phototaxis is generally also not inhibited. The ATP-level has been measured for *Chlorella* in the light and in the dark in the absence or presence of 10 μM *diuron* [65]. No differences were found.

Besides photosynthesis, other plant metabolic processes may be inhibited by *diuron* or related photosynthesis inhibitors, either directly or indirectly. Direct effects usually occur only at higher concentrations, compared with those needed to inhibit photosynthesis. Special emphasis has been laid on respiration studies with isolated mitochondria [66]. Studies with intact roots of *Allium sativum* showed a pronounced sensitivity of this non-chlorophyllous tissue towards higher concentrations of *diuron* and related herbicidal and non-herbicidal substituted ureas. The effects were observed on an ultrastructural level [67]. Interesting as these data are, they will only be of importance to the physiologist employing *diuron* in physiological experiments and probably bear little relevance to *diuron* toxicity during agricultural use. In the yeast *Saccharomyces cerevisiae* the site of respiratory inhibition by *diuron* is located between the cytochromes b and c, somewhat similar to antimycin [68, 69].

Fenuron [19, 40, 70] is a herbicide with a rather low pI_{50}-value of about 5.0. However, this shortcoming is obviously compensated for by a very good solubility in water (Fig. 17). Thus, e.g. in the inhibition of marine algae [71], the activity of *fenuron* is very low when compared to *diuron*, *neburon*, and *monuron*. In higher plants, however, where accumulation may occur with the transpiratory flow of water through the plant, *fenuron* activity compares more favourably [60].

Fluometuron [37] is a herbicide with two clearly independent mechanisms of action in plant metabolism. As a substituted phenylurea compound it exhibits a good photosynthesis-inhibiting activity. Moreover, it inhibits carotene biosynthesis in germinating seedlings and new growth, giving rise to white tissue [72–74]. *Fluometuron* shares this type of action with other herbicidal structures known as "bleaching herbicides" (p. 99). In *Abutilon theophrasti* germination, *fluometuron* increases the chloroplast granal size but decreases the chlorophyll content progressively with time [73]. Chlorophyll formation is apparently not inhibited directly but the pigment is destroyed in the light because of the missing photooxidative "carotene shield". In the unicellular algae *Chlorella pyrenoidosa* and *Euglena gracilis* this could be demonstrated by normal chlorophyll formation in the presence of *fluometuron* in the dark but not in the light [72].

Other studies have shown that at a concentration of 0.4 mM *fluometuron* has no effect on and is not bound by isolated protoplasts from tomatoes [38], and that it strongly uncouples oxidative phosphorylation with malate as a substrate in soybean mitochondria at 10–100 μM [75]. To sum up, it can be stated that the herbicide *fluometuron* specifically inhibits several plant metabolic processes, depending on the concentration and on the physiological stage of the plant.

Selectivity of *fluometuron* among crop plants and weeds, as is common with xylem mobile herbicides, is the result of more rapid detoxification in the resistant plant [76].

Isoproturon [37].

Linuron [37,40] is, among other crop plants, used for weed control in carrots. For this reason the influence of *linuron* on the carotene levels has extensively been investigated. Either an increased formation of α-carotene [77] or a delayed formation of β-carotene has been reported [78]. No pronounced concentration change was detected for other plant constituents like amino acids and ascorbic acid or the seed oil of soybeans [41]. However, the level of soluble sugars including sucrose may be decreased [79], undoubtedly as a consequense of partial and/or temporary photosynthesis inhibition by this herbicide [60]. There are two papers stressing the importance of *linuron* uptake into and transport within the plant for selectivity. Uptake of *linuron* by tomatoes via the roots and transport into the leaves is proportional to the transpiration rate [80] whereas, conversely, tolerance in umbelliferous plants (carrots, parsnip) has been reported to arise from a low transport rate into the shoot [81]. *Linuron* metabolism proceeds only slowly in resistant as well as in susceptible plants. The mode of toxic action is the same as for the other urea herbicides [82].

Methabenzthiazuron [30] and its relevant crop plant spring wheat have extensively been studied as a model system in the investigation of sublethal (physiological) effects of photosynthesis-inhibiting herbicides. Structure-activity correlations have shown that, in contrast to the phenylurea herbicides [134], in the benzthiazolylurea series a free -NH- group is not required for the inhibition of photosynthetic electron transport. Instead, the free electron pair on the urea nitrogen is thought to participate in herbicide binding at the site of action.

Methazole is metabolically activated in the treated plant to yield N-methyl-N′-dichlorophenylurea [83–85]. Both *methazole* and its primary metabolite have been reported to be phytotoxic. Selectivity arises from further metabolism [86].

Metobromuron [37].

For *metoxuron* [37], selectivity among different wheat varieties has been traced to different rates of detoxification by demethylation and further derivatization [43].

Meturin [87].

Monolinuron [37,47] has been checked, along with other urea herbicides, for the possible induction of chromosome aberrations in barley [88]. Although *monolinuron* was the most active compound of those investigated, the high concentrations required (250 to 1,000 $mg\,l^{-1}$) do not allow further conclusions to be made.

As with other photosynthesis-inhibiting herbicides, sublethal and physiological effects have also been observed with *monuron* [19, 37, 70, 89]. In *Phaseolus vulgaris*, an increase of dry weight production and of transpiration was detected [90]. As for other phenylurea herbicides, also *monuron* is taken up by and transported in the plant with the transpiration stream [90]. A differential sensitivity of CO_2-fixation in C-3 and C-4 plants (cf. p. 20) towards *monuron* has been reported by Malakondaiah and Fang [91]. The isolated mesophyll cells of C-4 plants, whose CO_2-fixation showed a decreased sensitivity towards *monuron*, may however not be relevant to the overall CO_2-fixation in the C-4 plant because in the intact leaf they are metabolically integrated with the bundle sheath tissue, especially with respect to their CO_2-fixation. Significantly, oxygen evolution was very sensitive to 0.3–0.6 μM *monuron* in all the cell types tested. It is at present not clear whether the mechanism of CO_2-fixation (C-3 vs. C-4 pathway) can provide a basis for selective herbicidal activity (cf. pp. 35 and 66).

An inhibition of respiration by *monuron* in the 0.1–1.0 mM concentration range has been observed with *Saccharomyces cerevisiae* [68]. The site of inhibition was localized between cytochromes b and c and was apparently the same as for *diuron*.

The decreasing effect of *neburon* [40, 92], *diuron*, *fenuron*, and *monuron* on the carbohydrates of marine algae has been studied [71, 93]. Only unicellular planctontic species have been tested, and the differences in sensitivity among several species in some cases were as high as 1:10. The concentrations employed ranged from 1 ppb for *diuron* to 1,000 ppb for *fenuron*, reflecting (a) differences in pI_{50}-values and (b) differences in water-solubility, with low solubility causing accumulation in the algal cell. The influence of *neburon* on the respiration in potato mitochondria has been studied [94, 95].

Noruron [40].

Siduron [40] is a weak photosynthesis inhibitor. It may act on another metabolic pathway besides photosynthesis because root growth is more sensitive than shoot growth in *Digitaria* and *Hordeum*, and because the inhibitory action on root growth is in part compensated by kinetin [96]. As for *neburon*, the effect of *siduron* on mitochondria has recently been studied [94].

Tebuthiuron is sometimes incorrectly called *butathiuron*. Furthermore, it should not be mistaken for N,N-dimethyl-N′-(5-t-butyl-1,3,4-thiadiazol-2-yl)urea, a compound which is more rapidly demethylated/inactivated in the resistant barley than in the more sensitive wheat plants [97]. The exact site of inhibitory action of *tebuthiuron* in chloroplast electron transport has recently been studied [35].

c. Carboxanilides

The herbicides of the anilide group (Table 5) only contain part of the urea structure but are equally good inhibitors of photosynthesis.

In a study by Alsop and Moreland [19, 40], *chlorancryl* (*dicryl*, WSSA) has been classified as an "inhibitory uncoupler" of photosynthesis together with *CIPC*, *dinoseb*, *ioxynil*, *oryzalin*, and *propanil* [98]. The action of these and other herbicidal structures on the light-activated, Mg^{2+}-dependent ATPase from chloroplasts was studied. These results should be integrated with other information when discussing the cause of toxicity after application of photosystem II-inhibiting herbicides (p. 66).

Cypromid [40].

Karsil [19, 40].

Pentanochlor [99].

Propanil [19, 40, 47, 98, 100, 101] is a herbicide which is extensively used for weed control in rice. Although it is a powerful inhibitor of photosynthetic electron flow at the "diuron site", at high concentrations other pathways also become in-

Table 5. Structures, water solubilities, and pI_{50}-values of photosynthesis-inhibiting anilide herbicides (anilides of carboxylic acids) (cf. Table 4)

Common name	Structure	Water sol. (mg l^{-1})	pI_{50}
Chlorancryl	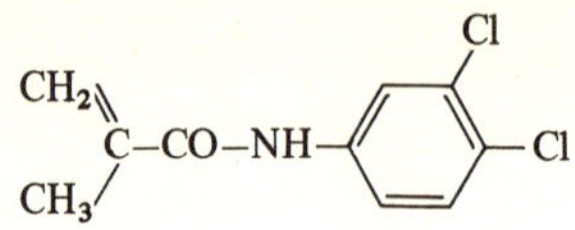	8	6.5–6.7
Cypromid	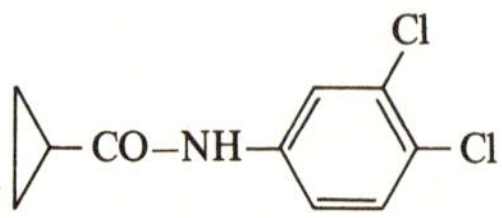	100	6.2
Karsil	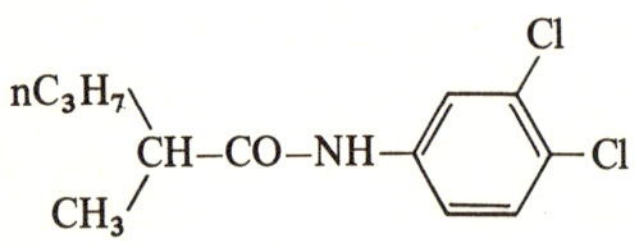	–	6.5–7.0
Pentanochlor	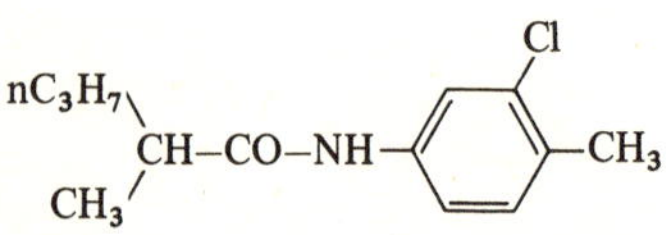	9	–
Propanil	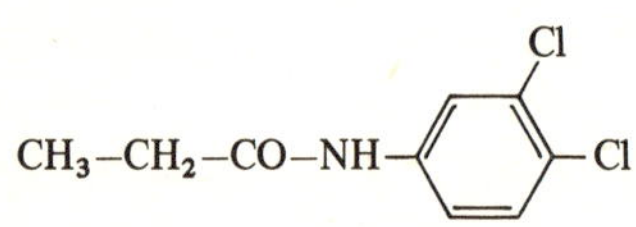	225	5.6–6.8
Swep	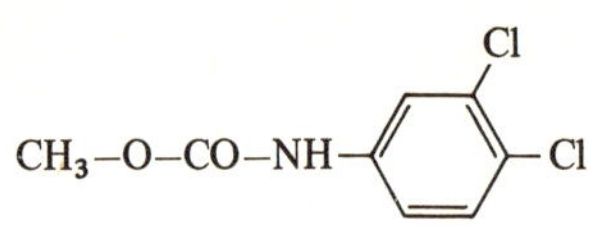	600	5.4–5.5

hibited. With 0.1 mM *propanil*, cyclic photophosphorylation is adversely affected [98, 102], and several types of plant membranes including those of chloroplasts and of red beet are damaged [103]. Oxidative phosphorylation is uncoupled at 0.1 mM *propanil* in *Ricinus communis* mitochondria [66], and after application of 0.1 mM *propanil* the ATP-level in soybean hypocotyls is severely decreased [104].

The resistance of rice plants towards *propanil* has been traced to an enzyme "aryl-acylamidase" which rapidly hydrolyzes the herbicide [105–110]. The enzyme has been characterized in vitro (k_m 25 μM, opt. pH 8.2, opt. temp. 50 C), and the substrate specificity has been determined. Interestingly, the enzyme may be inhibited by carbamate insecticides in vivo as well as in vitro leading to increased *propanil* toxicity [111].

Swep [19, 40, 47, 98, 100, 101] is a weak photosynthesis inhibitor. Basically, the molecule may not only be seen as an anilide like the other representatives of this group but may also be regarded as a carbamate (Table 39). However, while the typical carbamates are germination inhibitors, *swep* behaves more like an inhibitor of photosynthesis. Comparing the effect of 13 herbicides on lipid metabolism in *Chlorella*, Sumida and Ueda [112] found that *swep*, like other inhibitors of photosystem II, severely inhibited the biosynthesis of complex lipids from U-^{14}C-acetate. Thiolcarbamates and *2,4-D* did not reveal an inhibitory effect.

d. *s*-Triazine Herbicides

s-Triazine herbicides (Table 6) represent one of the economically most important groups of herbicides. Detected in 1955 at Geigy AG they have evolved, among others, into the economically and quantitatively most important herbicide, *atrazine*. In effect, *atrazine* became a herbicide of paramount importance because of its superior tolerance in maize. Some individual aspects of the different triazine herbicides will now be considered. General aspects are considered elsewhere (for the cause of toxicity see p. 64, for sublethal effects p. 70, for tolerance from site-of-action-change by mutation p. 55).

Among the *s*-triazines, *ametryne* [113, 114] and *prometryne* exhibit the strongest inhibitory effect not only on photosynthetic electron transport but interestingly also on cyclic photophosphorylation, mitochondrial state-3 respiration (corn, bean), hypocotyl hook opening in cotton, stimulation of germination in *Lactuca*, and, accordingly, phytotoxicity [114–116]. However, it is doubtful whether other inhibitions besides photosynthesis inhibition contribute significantly to the herbicidal action in the field. *Ametryne*, like other *s*-triazines, also displays an inhibitory effect on mitochondria from rat-liver, but only at concentrations considerably above 0.1 mM.

Atraton [113, 114].

Several studies have confirmed that the binding site in chloroplasts for *atrazine* [19, 20] is identical with the site for *diuron*-binding [24]. Inhibition of CO_2-fixation in vivo by *atrazine*, as is also the case with other photosynthesis inhibitors, leads to stomata closure because of the rising internal CO_2-level [117, 118]. Significantly, stomata closure is observed in CO_2-free air. Photorespiration was inhibited whereas the oxidation of exogenously applied glycolate was unaltered. Therefore, no direct inhibition of photorespiration by *atrazine* can be deduced. However, the inability of *atrazine* and other triazines to adequately control plants with C-4 metabolism (e.g. *Panicum miliaceum, Digitaria sanguinalis, Cyperus rotundus, Zea mays*) (p. 20) could indicate that photorespiration is important for the selectivity detected with these herbicides. Starting from this fact, and taking into consideration that CO_2 deficiency and *s*-triazine herbicides cause the same type of damage, it has been argued [119] that the depletion of carbon pools by photorespiration and metabolite transport out of the inhibited leaf could eventually lead to cell damage by uncontrolled photorespiration/photooxidations. This assumption would basically also apply to other herbicides inhibiting photosynthesis. Indeed, "tropical grasses" with the C_4-type CO_2-fixation pathway are generally hard to control with

Table 6. Structures, water solubilities, and pI_{50}-values of *s*-triazine herbicides (cf. Table 4)

Common name	Structure	Water sol. ($mg\ l^{-1}$)	pI_{50}
Ametryne	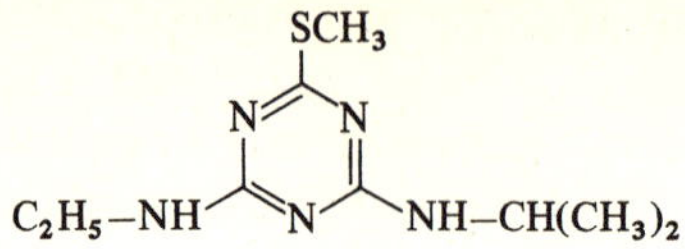	193	6.8–7.3
Atraton	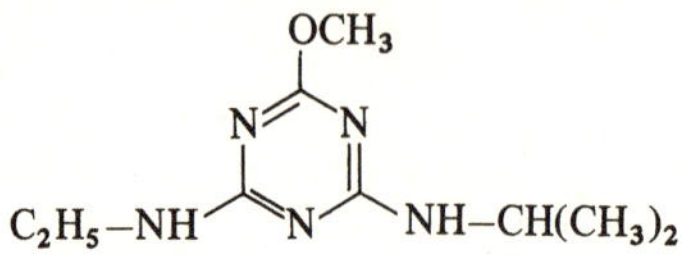	1,800	6.1–6.3
Atrazine	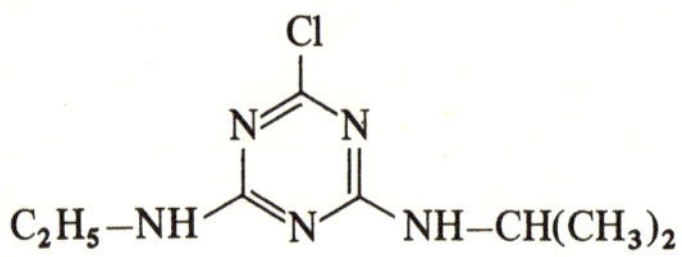	33 (70)	6.1–6.6
Cyanazine	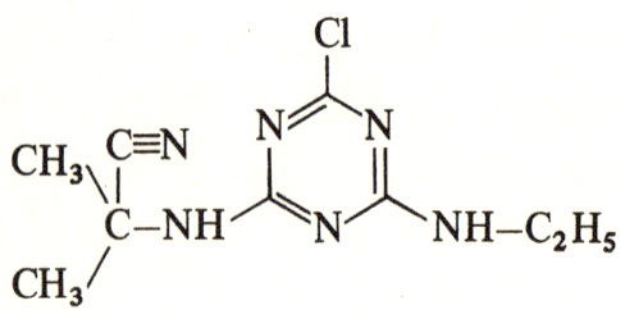	171	–
Prometon	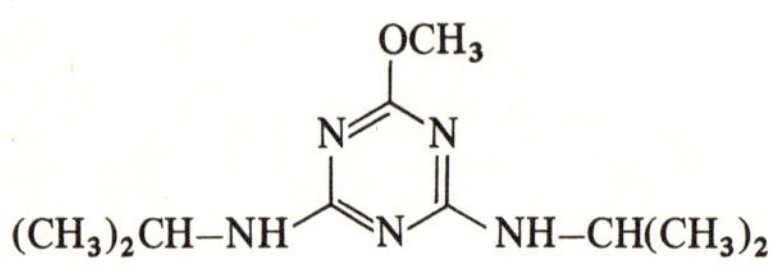	750	5.9
Prometryne	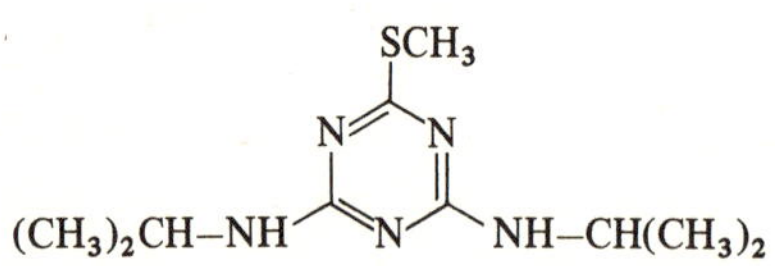	48	7.0
Propazine	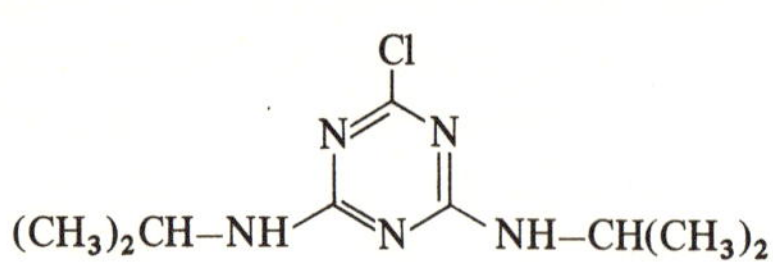	8.6	5.4–6.3
Simazine	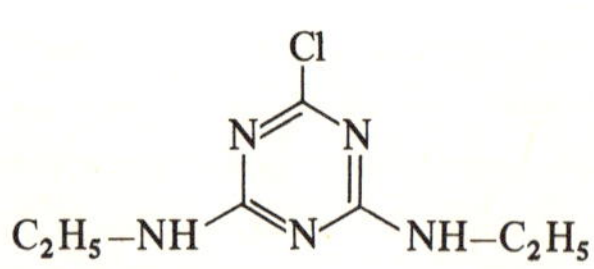	5	5.4–6.4

Table 6 (continued)

Common name	Structure	Water sol. ($mg\ l^{-1}$)	pI_{50}
Simeton	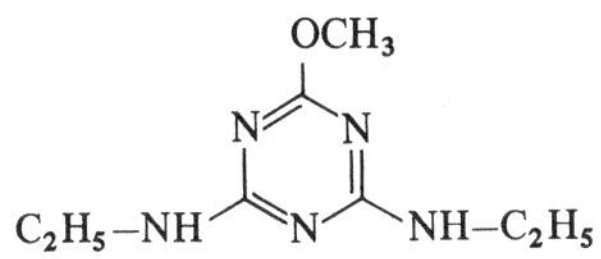	2,500	5.2–6.3
Simetryne	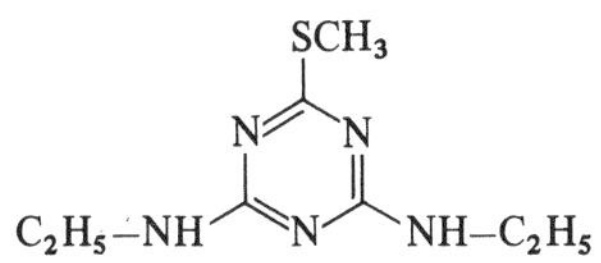	450	6.6
Terbutryne	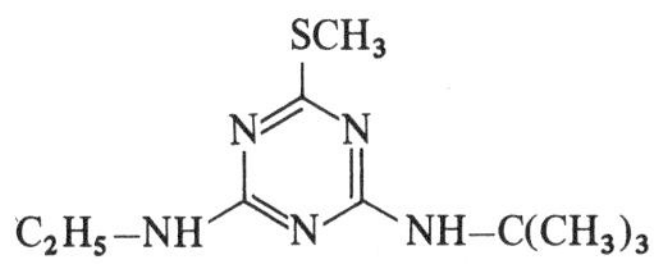	58	–
Trietazine	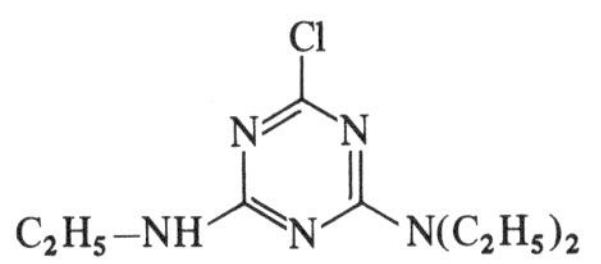	20	3.6–4.7

any type of herbicide. However, some photosynthesis inhibitors (e.g. *metribuzin*) are toxic towards a number of "tropical grasses" which do not display photorespiration. Moreover, in plants which have been preadapted to low light intensities, herbicidal inhibition upon exposure to light of high intensity leads to very rapid damage (cf. p. 65), which cannot easily be explained by action via photorespiration. In addition, toxicity symptoms in the C-4 plant *Echinochloa crus-galli* (Barnyard grass) may already be seen on an ultrastructural level (for further details see p. 66) 2 h after application of 2–20 ppm *atrazine* [120]. The organelle damaged first is the chloroplast, more precisely the chloroplast membrane. Possibly, then, under specific climatic conditions, the existence of a photorespiratory pathway might contribute to a more rapid and more pronounced expression of phytotoxicity.

In addition to the possible involvement of photorespiration in the generation of phytotoxicity, there exists a true physiological resistance by selective detoxification in corn plants [121–125]. The initial detoxifying metabolic step is the formation of a glutathione conjugate in the resistant plant. The reaction is catalyzed by a cytoplasmic glutathione-S-transferase [126]. Furthermore, both in susceptible and in resistent plants a non-enzymatic degradation catalyzed by benzoxazin occurs. In addition to these "traditional" selectivity mechanisms, resistant mutants which possess a herbicide-resistant electron transport system have recently been observed (p. 55)

Atrazine was assumed to induce chromosome aberrations at normal field rates in sorghum and to initiate the emergence of polyploidic weeds (e.g. giant foxtail) [127]. However, these results could not be confirmed by other workers [128]. Also, the incorporation of *atrazine* into the nucleic acids of *Escherichia coli* could not be substantiated [129]. On the other hand, aqueous extracts from *atrazine*-treated corn contained a mutagen active in yeast [130]. The mutagen was suspected to be an *atrazine* metabolite. *Atrazine* promoted the RNA-synthesis with isolated soybean (but not corn) chromatin when added before the chromatin extraction [131]. This increase might indicate a growth-hormone type activity attributable to many s-triazine herbicides (p. 76).

An inhibitory activity on electron-transport in cucumber mitochondria may be seen at concentrations $>10\ \mu M$ *atrazine* [132].

Many experiments have been performed with *atrazine* demonstrating the action of sublethal concentrations on plant metabolism and composition, and on the inherent growth hormone activity in relation to other *s*-triazines. These results will be discussed elsewhere (p. 70).

The rate of recovery of CO_2-fixation from initial inhibition by *atrazine*, *cyanazine* or *cyprazine* has been measured in single leaves of nine grasses [133]. Whereas in most cases the rate of recovery of CO_2-fixation reflected rapid rates of detoxification in the leaves of tolerant plants, some species, e.g. *Digitaria sanguinalis* and *Panicum miliaceum* treated with *cyanazine*, demonstrated considerable tolerance in spite of low recovery rates from inhibition of CO_2-fixation. Consequently, other factors than foliar detoxification may play an important role in the tolerance of some grasses to *cyanazine*.

Prometon [135].

Prometryne [116] is one of the most active inhibitors of photosynthesis in the *s*-triazine series both in vitro and in vivo [136]. Whereas the carbon-fixation pattern after feeding with $^{14}CO_2$ is unchanged in the dark, in the light there is an increased labelling of aspartate and organic acids but a decreased labelling of amino acids, sugars, and glycerate. Of course, the share of dark CO_2-fixation products in the light in inhibited photosynthetic tissue should increase in short-time experiments. However, after a few days of adaptation of the plant to a decreased rate of photosynthetic CO_2-fixation, the flow of originally fixed carbon into different metabolic pathways might change completely [137].

In addition to the inhibition of photosynthesis, *prometryne* at elevated concentrations also appears to exert a profound influence on respiration [75, 114, 115, 138]. Mitochondria from soybean hypocotyls were sensitive to 0.1 mM *prometryne* in state 3. The inhibition could be releaved by the uncoupler 2,4-dinitrophenol, which indicates an inhibition of energy transfer. Respiration in *Chlorella* was increased by *prometryne* and *diuron*, thus indicating an uncoupling activity of these herbicides in vivo in this alga.

In a comparison of the influence of different *s*-triazine herbicides on electron transport in isolated chloroplasts and mitochondria, *prometryne* was the most active compound (Table 7). Interestingly, the inhibition was nearly equal in plant and in animal mitochondria. An influence of *prometryne* on mitochondria was also revealed by earlier studies where the effects on photosynthesis and nucleic acid synthesis in spinach leaves where compared kinetically [139]. The same conclusion

Table 7. Influence of *s*-triazine herbicides on the activities of isolated chloroplasts and mitochondria. Respiration was tested with 0.1 mM, photophosphorylation with 1 μM herbicide [114]. Least significant differences between 5 and 10%.

Treatment	% Inhibition of			pI_{50} PS inhibition
	Respiration in mitochondria		Cyclic photophosphorylation	
	From bean hypocotyl	From rat liver		
Prometryne	25	30	52	6.95
Ametryne	19	20	18	6.8
Simetryne	9	10	8	6.55
Atrazine	8	10	9	6.4
Atraton	13	12	34	6.3
Simeton	8	4	15	6.3
Prometon	14	16	28	5.9

could be drawn for *ioxynil*, a well-known respiratory uncoupler and photosynthesis inhibitor which was included in the experiments. An influence of *prometryne* on RNA synthesis in *Phaseolus* leaf discs has not been found (Whitenberg et al. [137]). However, *prometryne* or a degradation product thereof instead of uracil and thymidine became incorporated into *Escherichia coli* nucleic acids [140].

Prometryne has been shown to be more mobile in sensitive plants, an observation which might be relevant for selectivity considerations [141].

Propazine [19, 20].

The inhibition of the CO_2-fixation by *simazine* [19, 20, 100] has extensively been studied [60, 142, 143]. Alternatively, the inhibition of oxygen evolution can be measured [144]. The CO_2-fixation recovers rapidly from *simazine* inhibition in resistant maize plants but slowly in sensitive sugar beet (Fig. 19). The reason for this rapid recovery is the conversion of *simazine* (as in the case of *atrazine*) in the resistant plants into S-glutamylhydroxysimazinylcystein via hydroxysimazine [145, 146]. Another mechanism of selectivity, recently established, postulates a difference at the binding site resulting from mutation and selection (p. 55).

A comparison of 13 herbicides with respect to their inhibition of ^{14}C-acetate incorporation into complex lipids revealed that in Chlorella, *simazine* was the most active compound [112]. In general, only photosystem II-inhibitors exhibited an inhibitory effect in these experiments. The relative incorporation into different lipid fractions was also changed, with the mono- and digalactolipid fraction decreasing.

An interesting effect is the bleaching of green leaves after treatment with *simazine* [147, 148]. This type of bleaching should not be mistaken for chlorosis in newly developing tissue caused by inhibitors of carotenoid biosynthesis (p. 99). Whereas the former produces yellowish tissue, true chlorotic tissue appears pale white. Bleaching after application of *simazine* is most clearly observed close to the leaf veins, presumably because the concentration there is highest, and appears late after initial photosynthesis inhibition. It is typical of slowly acting photosynthesis inhibiting herbicides. In isolated chloroplasts, *simazine* as well as *atrazine* induce

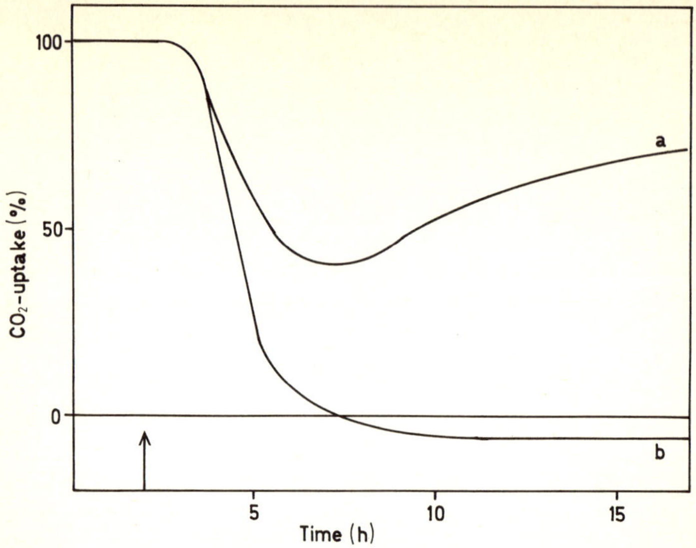

Fig. 19. Effect of *simazine* on CO_2-fixation in resistant maize *a* and in sensitive sugar-beet *b*. 20 μM of *simazine* was given via the nutrient solution at ↑ [60]

slight variations of unsaturated fatty acids which may, however, be insignificant at lower concentrations [149].

Sublethal effects of *simazine* on crop plant composition and metabolism have very often been reported. In fact, *simazine* is one of the compounds which has stimulated the interest in sublethal effects (for more details see p. 70).

Simeton [19,47] and *isocil* inhibited photosynthesis in intact sugar beet and bean plants to a greater degree than transpiration [59, 150, 151]. Sensitivity was highest when the plants were exposed to low-intensity light before and exposed to intense light after treatment with the inhibitor. (For an explanation for this phenomenon see p. 65). Another interesting result showed that leaf damage was comparable after *simeton* treatment or in CO_2-free air when the plants were irradiated. This observation has important implications for our understanding of the mechanism of toxicity in herbicidally photosynthesis-inhibited plants (p. 68). From his experiments, which were extended to CO_2-free nitrogen and which were conducted both in the light and in the dark with C-3 and C-4 plants, van Oorschot concluded that photorespiration did not contribute to phytotoxicity, neither in CO_2-free air nor during herbicidal action. This is in contrast with other considerations discussing the involvement of photorespiration in herbicide toxicity via inhibition of photosynthesis (p. 35). The currently available evidence does not seem to support involvement of photorespiration in the generation of phytotoxicity.

Simetryne [47, 114].

Investigations into the mechanism of selectivity of *terbutryne* among cotton (resistant) and bean (sensitive) have shown the involvement of both detoxification and uptake and/or transport of the herbicide [152, 153]. Similarly, the differential toxicity of *terbutryne* and *prometryne* in peas was attributed to differences in soil availability, rate of transport and detoxification [154].

The importance of ecological factors, including light, humidity and temperature, for the herbicidal activity has also been confirmed for the action of *terbutryne* [155,156].

Trietazine [19,20].

e. *as*-Triazinone Herbicides

Three compounds of the *as*-triazinone group (Table 8) of herbicides have gained commercial importance so far. Of these, *isomethiozin* is a derivative of *metribuzin* which is metabolized into the active herbicidal structure in the sensitive plant. Free *metribuzin* can therefore be found in *isomethiozin* treated tissue.

Metamitron [157,158] is a fairly weak inhibitor of photosynthesis but has gained importance because of its high tolerance in sugar beet. The compound is only mobile in the xylem [159] and is rapidly metabolized in the tolerant beet leaves [158,160]. The initial detoxifying reaction involves deamination of the molecule which also occurs in UV light [161]. Extensive studies on the deaminating enzyme from sugar beet leaves have revealed that this enzyme uses part of a peroxisomal reductive electron transport chain which normally reduces oxygen but may also deaminate *metamitron* [162,163].

Metribuzin [164–166] is one of the most active herbicides known so far, only requiring an application of 350–700 g/ha of the active compound. This high activity is due to the high water solubility and a suitable pI_{50}-value (Fig. 17), but additional factors, e.g. the behaviour in the soil, may be important as well [167]. *Metribuzin* inhibits the photosynthetic electron transport at the same site as *diuron*

Table 8. Structures, water solubilities, and pI_{50}-values of *as*-triazinone herbicides (cf. Table 4)

Common name	Structure	Water sol. ($mg\ l^{-1}$)	pI_{50}
Isomethiozin	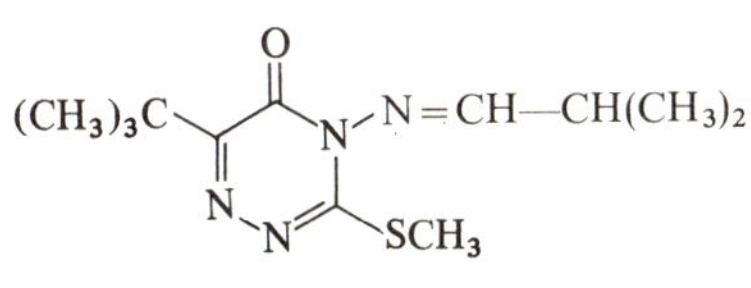	1,000	<5
Metamitron	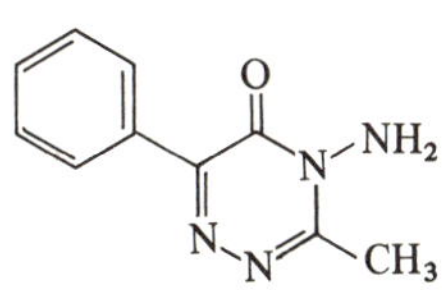	1,800	5.4–6.1
Metribuzin	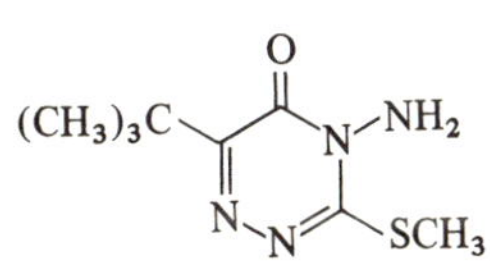	1,200	6.5–6.7

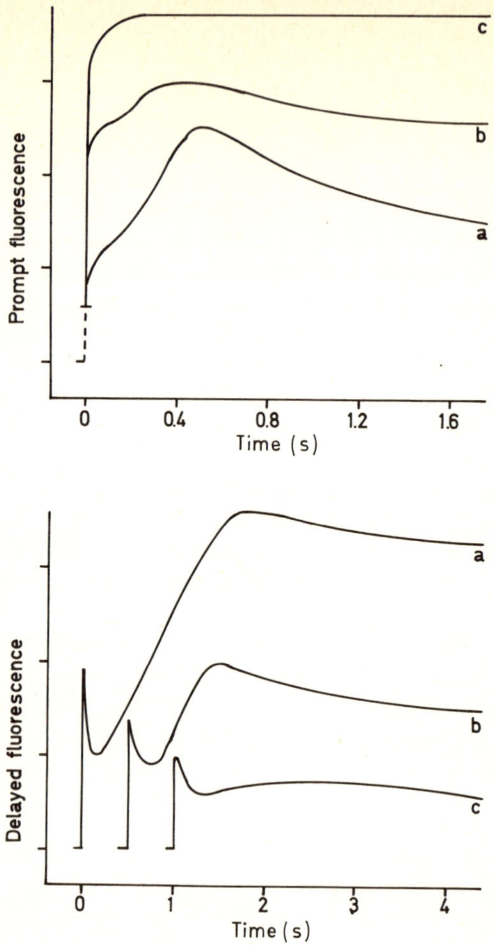

Fig. 20. Prompt and delayed fluorescence in *Chlorella pyrenoidosa* after the addition of *metribuzin*. *a* control, *b metribuzin 0.5* μM, *c metribuzin* 5 μM [166]

[166] and also binds to the same site [21]. The fluorescence yield is stimulated by metribuzin to a maximum in the same way as with *diuron* (Fig. 20 and p. 21). The delayed fluorescence, which is measured after switching off the light, is similarly decreased by both herbicides. Cyclic photophosphorylation with PMS as a cofactor is only slightly inhibited by 15 μM *metribuzin*.

The toxicity symptoms and the conditions for phytotoxicity are the same as with other photosynthetic inhibitors, especially as discussed for *simazine* (p. 39) and *simeton* (p. 40). Specifically, exposure of the plants to weak light before and intense light after treatment with the inhibitor enhances the herbicidal activity [168–170]. The toxicity became visually expressed as necroses and bleaching along the leaf margins and leaf veins and sometimes in the intercostal areas [171]. Slight changes in the oil composition of soybean have also been reported [172].

Many studies were concerned with the elucidation of *metribuzin* tolerance among different plants. *Metribuzin* tolerance results from both physiological mechanisms [173–179] and methods of agricultural practice [180]. A more rapid detoxification of *metribuzin* in the tolerant plant species or variety has commonly been found to be the most prominent factor in plant tolerance, but differences in the uptake and/or transport rates may in some cases also contribute to the observed tolerance. More extensive analyses of the breakdown of *metribuzin* in vivo and in vitro in sensitive and resistant plants have not yet provided sufficient information for a general understanding of the enzymatic degradative processes [163, 177, 181, 182].

f. Uracile Herbicides

Only few studies have been devoted to the inhibiting effect of uracile herbicides (Table 9) in photosynthesis, and even less studies have concentrated on individual uracile compounds. However, the effects of uracile herbicide action closely resemble those caused by other inhibitors of photosynthesis.

Table 9. Structures, water solubilities, and pI_{50}-values of uracile herbicides (cf. Table 4)

Common name	Structure	Water sol. ($mg\ l^{-1}$)	pI_{50}
Bromacil	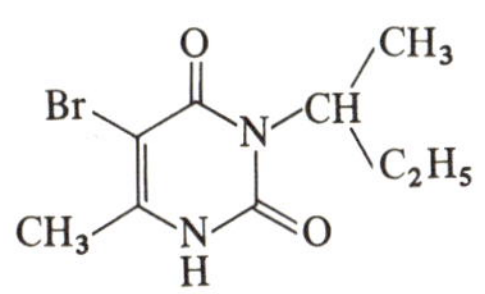	815	5.8–6.3
Isocil	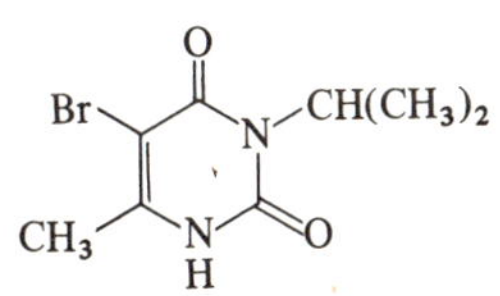	2,150	6.0
Lenacil	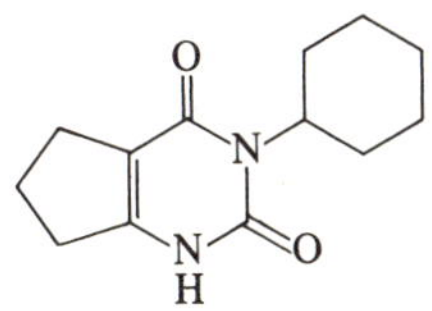	6	6.6
Terbacil	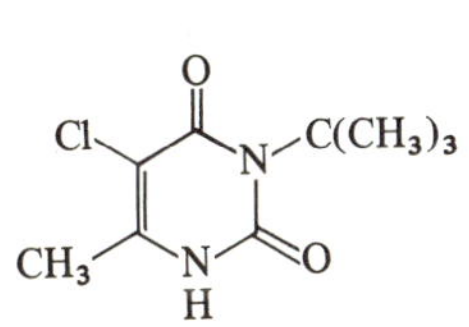	710	6.4

Bromacil [40, 185], like *fluometuron* and *atrazine*, but at lower concentrations, retards the senescence in maize leaves [186]. This effect may be seen with all inhibitors of photosynthesis and has been interpreted to be a consequence of partial inhibition of photosynthesis.

The reason for the selective action of *bromacil* has been studied by comparing *Dactylis glomerata* (sens.) and *Poa pratense* (res.) [187]. Photosynthesis was initially inhibited in both types of plants, but only in the resistant species was CO_2-fixation recovered after 6 days. Uptake and transport of the inhibitor were correlated with the transpiration rate, but the detoxification rate was considerably higher in the resistant plants.

Structure-activity correlations have been reported for the bromacil-related series of N-3-alkoxyuracil derivatives [303].

Isocil [40] has been included in studies with intact plants [59]. As with the other herbicides studied, *isocil* was also shown to reduce photosynthetic CO_2-fixation more strongly than transpiration.

Lenacil [40].

Terbacil has been reported to exhibit pesticidal interference [185]: the insecticide fonofos simultaneously increases the rate of uptake and decreases the rate of detoxification of this herbicide in *Mentha piperita*, thereby increasing the phytotoxicity [188].

g. Biscarbamates

Of the three herbicides grouped as biscarbamates (Table 10), *karbitulate* [46] can also be classified as an urea compound. However, by omitting the carbamoyl group of the molecule the even weaker inhibitor of photosynthesis, *fenuron*, is obtained. Without the urea moiety of the molecule the herbicidal activity is completely lost.

Desmedipham [102] shows the usual characteristics of photosynthetic inhibitors, like e.g. increased phytotoxicity at high light intensities [189] and the dependence of phytotoxicity on climatic conditions [190]. Bethlenfalvay et al. also reported on the enhancement of proton translocation in a cyclic system in isolated spinach chloroplast fragments by *desmedipham*, *diuron*, and *propanil* (Fig. 21). These results clearly show that *desmedipham* acts in the same way as *diuron* and *propanil*. The enhancement of proton translocation, which is equivalent to an increase of cyclic photophosphorylation (pp. 21, 114) becomes possible because the open electron transport is cut off by these herbicides, and the possible redox interference from photosystem II is eliminated (Figs. 14, 15). Photosystem I, which catalyzes cyclic photophosphorylation, is obviously not inhibited by the inhibitor concentrations employed (6 μM).

The inhibition of photosynthetic electron transport by *phenmedipham* [191–193] has been clearly documented both in vivo [31] and in vitro. The binding site in the chloroplast is identical with the site for *metribuzin* and *atrazine* [21]. Structure-activity correlations have revealed that a free amino hydrogen is required at the -NH-CO-O-CH_3 group in the biscarbamate series for the inhibition of photosynthetic electron transport [191]. This is in contrast to the fact that in the benzthiazolylurea series a free electron pair on the urea nitrogen is obviously suf-

Table 10. Structures, water solubilities, and pI_{50}-values of biscarbamate herbicides (cf. Table 4)

Common name	Structure	Water sol. (mg l^{-1})	pI_{50}
Desmedipham	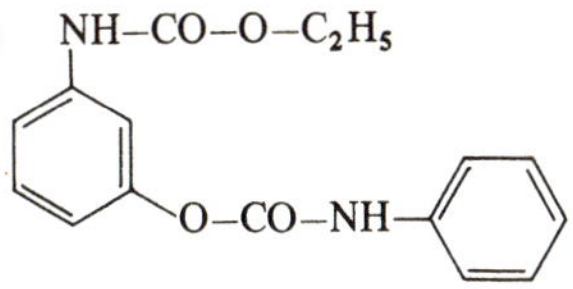	7	6.2
Karbutilate	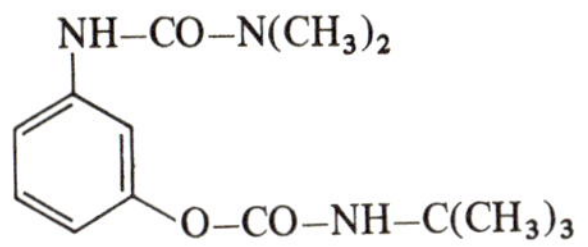	325	5.7
Phenmedipham	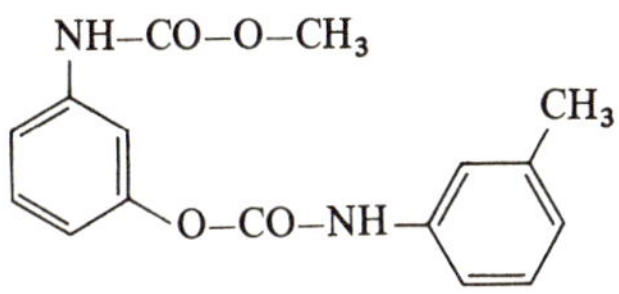	< 10	6.8–7.0

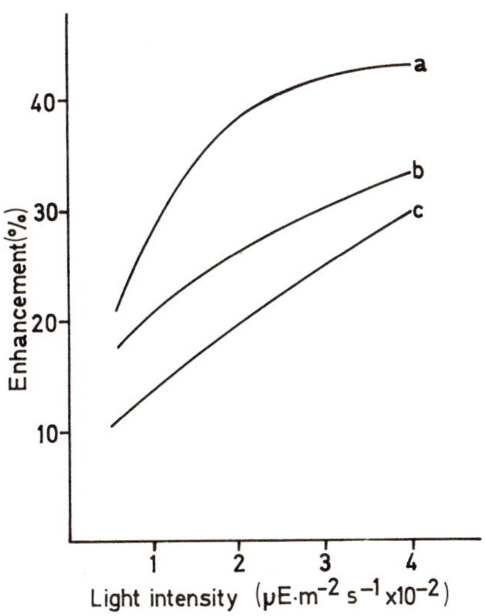

Fig. 21. Proton-translocation enhancement in spinach chloroplast fragments in a red light cyclic system. *a diuron, b desmedipham, c propanil;* each applied at a concentration of 6 μM [102]. Reproduced with permission of the Weed Science Society of America

ficient to cause inhibition (p. 32). Chloroplasts isolated from either sensitive spinach or resistant sugar beet leaves exhibited the same sensitivity towards *phenmedipham*. Here, selective herbicidal action is also due to rapid detoxification in the resistant plants [192], together with differences in the rate of uptake and rate of transport [194].

Respiratory electron transport in mitochondria from cucumber was not inhibited or uncoupled by 0.38 mM *phenmedipham* [193].

h. Pyridazinone Herbicides

Three different modes of action have been reported for pyridazinone herbicides (Table 11): inhibition of photosynthesis, inhibition of carotenoid biosynthesis and influence on lipid biosynthesis [195]. In spite of the narrow range of pI_{50}-values measured for the three pyridazinone herbicides described here, only *pyrazone* [30, 196, 197] appears to act in vivo as an inhibitor of photosynthesis. For *metflurazone* and *norflurazon* [30], inhibition of photosynthesis is not relevant during normal agricultural use; these herbicides are applied pre-emergence and give rise to chlorotic seedlings (p. 101), which of course do not possess photosynthetically active chloroplasts. Furthermore, *metflurazone* was transported out of the roots only to a very low extent and accumulated in the lysigenous glands of tolerant cotton plants [198, 199].

Pyrazone acts as an inhibitor of photosynthesis during normal agricultural use [200, 201]. Chloroplasts isolated from either tolerant sugar beet leaves or sensitive *Sinapis* and *Chenopodium* leaves did not differ in their pI_{50}-values [31, 196]. However, the herbicide is rapidly detoxified by the formation of an N-glucoside in sugar

Table 11. Structures, water solubilities, and pI_{50}-values of pyridazinone herbicides (cf. Table 4)

Common name	Structure	Water sol. ($mg\ l^{-1}$)	pI_{50}
Pyrazone	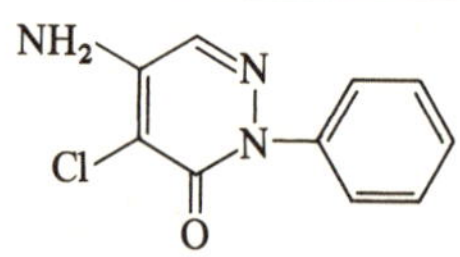	300	4.8–5.4
Metflurazone	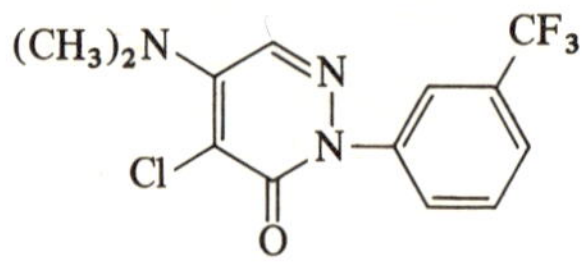	–	5.4
Norflurazon	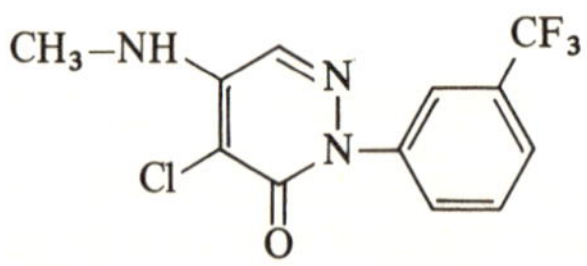	28	4.6

beet [202, 203], resulting in a rapid recovery of photosynthetic CO_2-fixation [60]. In vitro detoxification has been achieved in sugar beet but not in *Sinapis* homogenates [31]. It can therefore be assumed that *pyrazone* tolerance in sugar beet is the result of rapid detoxification in the plant.

Competition experiments with *atrazine* and other photosystem II inhibitors in isolated chloroplasts have confirmed the identity of the *pyrazone* binding site with the binding site for other photosystem II inhibitors [21]. Cyclic photophosphorylation with PMS as a cofactor could be sensitive to *pyrazone* [197]. Respiration, however, was not inhibited [196].

i. Hydroxybenzonitriles

The action of halogenated hydroxybenzonitrile herbicides (Table 12) may be described by two important mechanisms: inhibition of photosynthetic electron transport at the common photosystem II site [40, 98, 204] and inhibition [132] or uncoupling [60, 205] of oxidative phosphorylation and respiration.

Whereas *bromoxynil* [40, 205, 206] is a very potent inhibitor of photosynthetic electron transport, *chloroxynil* [40] is a rather poor inhibitor and has not gained commercial importance. The action of *Bromoxynil* on intact plants has been found to depend on climatic conditions in the usual manner (p. 65). The herbicide is rapidly detoxified in the resistant plants [208].

Table 12. Structures, water solubilities, and pI_{50}-values of hydroxybenzonitriles (cf. Table 4)

Common name	Structure	Water sol. ($mg\ l^{-1}$)	pI_{50}
Bromoxynil	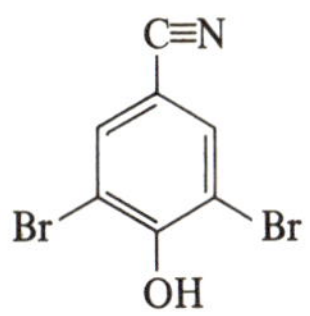	130	4.8–6.6
Chloroxynil	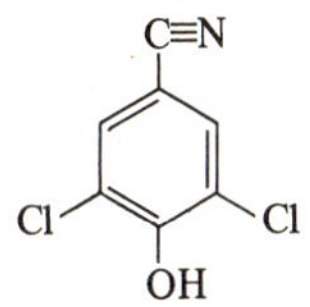	–	3.5
Ioxynil	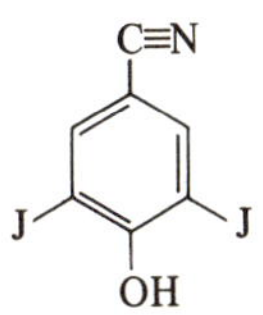	50	6.0–6.8

An interesting question is whether inhibition of photosynthesis or uncoupling of oxidative phosphorylation (p. 115) contributes more to the herbicidal activity of *ioxynil* [40, 204, 206]. Basically, this might depend on the method of application and on the physiological status of the tissue. After a leaf spray the main action could originate from the inhibition of photosynthesis [59]. However, when uptake of the herbicide occurs via the root, respiratory uncoupling seems to hinder the further uptake and transport into the leaf, and photosynthesis cannot be inhibited by more than 50%. A similar situation results when $^{14}CO_2$-fixation and ^{32}P-incorporation are compared: ^{32}P-incorporation is inhibited much more rapidly by *ioxynil* indicating a primary action of this herbicide on respiratory ATP-synthesis in these experiments [139]. Also, inhibition of mitotic processes is presumed to be a consequence of a decreased ATP-supply due to uncoupled oxidative phosphorylation [209].

Attention has been paid to DIHB (diiodohydroxybenzoic acid) which is an *ioxynil* metabolite in the soil and which causes growth regulatory effects comparable to other halogenated benzoic acids. DIHB inhibits ethylene biosynthesis and stimulates root elongation [210, 211, 211 a]. Since ethylene inhibits root growth, the stimulation by DIHB of root elongation could be a consequence of decreased ethylene concentrations.

k. Nitrophenols

Analogously to halogenated benzonitriles, nitrophenols (Table 13) may also act by two different mechanisms: inhibition of photosynthetic electron transport and uncoupling of oxidative phosphorylation. Today, these herbicides are only of limited practical importance because of their high mammalian toxicity. Since the applica-

Table 13. Structures, water solubilities, and pI_{50}-values of nitrophenols (cf. Table 4)

Common name	Structure	Water sol. (mg l^{-1})	pI_{50}
Dinoseb	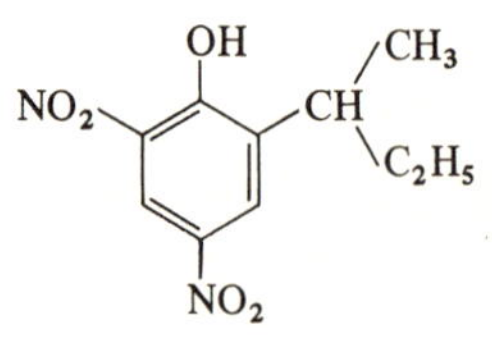	52	5.1–5.8
DNOC	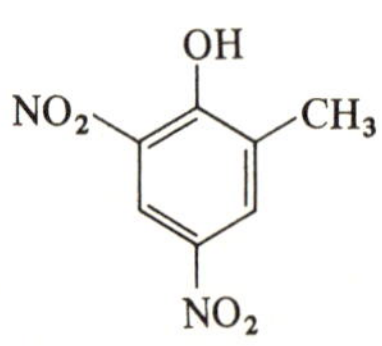	130	3.7–6.6

tion is as a post-emergence contact herbicide, both the action on photosynthesis and on respiration may contribute to the phytotoxicity under field conditions.

Because of the dual action on electron transport and phosphorylation, these compounds have also been termed "inhibitory uncouplers" [98, 212, 213]. The uncoupling activity is related to the ability of these compounds to transport protons through membranes, thereby decreasing transmembrane pH gradients (p. 114). The inhibitory action of *DNOC* on chloroplast electron transport has been found to be comparable with that of *diuron*. At the photosystem II site, a slight difference in the type of binding has been inferred. Similarly, recent studies with chloroplasts from *s*-triazine-resistant weeds (p. 55) have confirmed an own binding area to exist for nitrophenols which, however, overlaps with the binding areas of other photosystem II inhibitors.

The dual in vivo action of *DNOC* on photosynthesis and respiration has been studied by van Oorschot [59].

I. Benzimidazole Herbicides

In the benzimidazole herbicide series, similar to halogenated hydroxybenzonitriles and to nitrophenols (Tables 12–14), both photosynthetic electron transport and oxidative phosphorylation must be considered as sensitive metabolic pathways. Benzimidazoles and related compounds have been classified as "NH-acidic π-excessive heteroaromatica" [134], and their biological activity could be quantitively correlated with the pK_a-value and the partition coefficient. An acidic NH-group, which is part of an aromatic π-system, is necessary for biological activity. Certain members of the family have pI_{50}-values as high as 8.0 [214, 215], and in the alkylbenzimidazole series potent respiratory uncouplers are obtained [216]. The respiratory uncoupling activity of the compounds listed in Tables 13–15 is discussed separately (p. 115).

Fluromidine [217, 218].

Table 14. Structures, water solubilities, and pI_{50}-values of benzimidazole herbicides (cf. Table 4)

Common name	Structure	Water sol. (mg l^{-1})	pI_{50}
Chlorfurazole	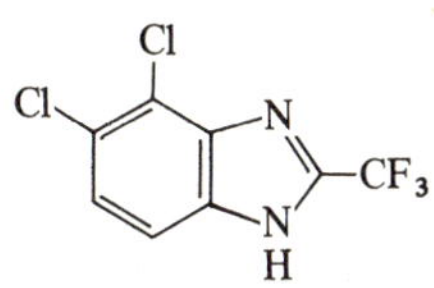	6	5.2
Fluromidine	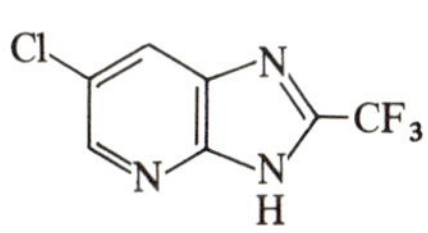	–	5.6–5.8

IV. Unclassified and Quinoid Photosynthesis-Inhibiting Herbicides

Several chemical structures among the established inhibitors of photosynthesis do not fit any of the well-known compound classes. Two of these herbicides are listed in Table 15. One of these, *bentazone*, is commercially important.

Bentazone [100, 219, 220] poses a special problem in the understanding of the inhibition of photosynthetic CO_2-fixation in vivo [221] because of the very poor pI_{50}-value. Several suggestions have been made which all could be of some relevance to the in vivo activity: (a) the high water-solubility leads to accumulation of the herbicide in the plant [100], (b) a metabolite of *bentazone* may act at a different herbicidal site [221], (c) high *bentazone* concentrations in specific compartments might block some other metabolic pathway in addition to photosynthesis, and (d) *bentazone* or a metabolite appears to be irreversibly bound to the chloroplast leading to its inactivation [222, 226].

Table 15. Structures, water solubilities, and pI_{50}-values of unclassified and quinoid photosynthesis-inhibiting herbicides (cf. Table 4)

Common name	Structure	Water sol. ($mg\ l^{-1}$)	pI_{50}
Bentazone	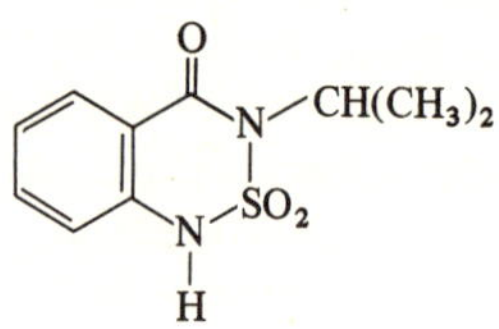	500	3.7–4.4
Ethofumesate	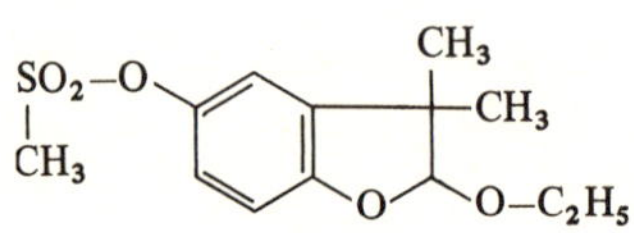	110	–
Chloranil	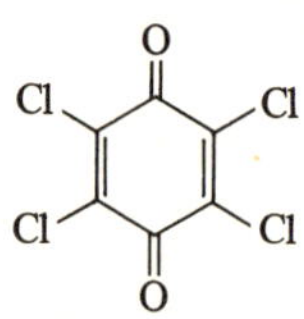	250	–
DBMIB	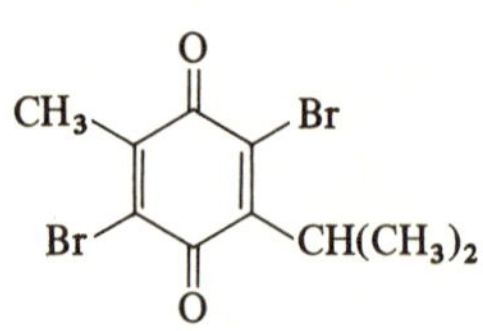	–	6.1
Dichlone	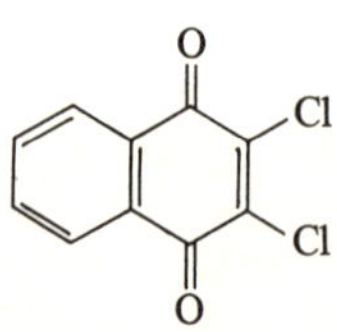	0.1	–

The very low pk_a-value of 3.5 means that *bentazone* is largely dissociated at physiological pH [223]. Whereas the anionic *bentazone* is readily xylem-mobile in the plant [224], the undissociated lipophilic compound, which is present in the equilibrium in only low amounts may, however, be the actual inhibitor. This would require that undissociated *bentazone* is trapped and slowly accumulates in the lipophilic chloroplast membrane system, thereby irreversibly blocking the electron transport [219, 223]. The rather slow accumulation would produce a poor pI_{50}-value in isolated chloroplasts but would lead to a better in vivo photosynthesis-inhibiting capacity. The pk_a-values for several other photosystem II-inhibitors have also been reported [225]. The site of inhibitory action of *bentazone* in the isolated chloroplast is the reducing side of photosystem II. Cyclic photophosphorylation is not inhibited.

The different tolerance of crop and weed plants towards *bentazone* reflects differences in uptake and detoxification of the herbicide. The chloroplasts are equally sensitive when isolated from tolerant or from resistant plants. Whereas Retzlaff et al. [226] have reported that tolerant plants always take up less herbicide, Mine et al. [227] found no differences in herbicide uptake and translocation. However, there exists general agreement that *bentazone* is more rapidly degraded in the resistant plant [228]. Degradation proceeds via the formation of 6- and 8-hydroxy-bentazone [229].

Ethofumesate [230].

In Table 15 three quinones are included which act on both mitochondrial and chloroplast electron transport systems. DBMIB specifically inhibits photosynthetic electron flow at the reducing side of the plastoquinone pool, as mentioned before (Figs. 14, 15; p. 24) [102, 231]. The structure-activity correlations have been studied [232]. However, DBMIB is also an inhibitor of the mitochondrial electron transport system [233]. The I_{50}-values reported are 20 μM for the cyanide-insensitive and 40 μM for the cyanide-sensitive respiration pathway.

Chloranil and *dichlone* have fairly positive redox potentials ($E'_0 = +482$ and $+463$ mV, respectively) [234]. They are both potent inhibitors of the photoassimilation of acetate and also of the fixation of CO_2 in *Chlorella* in the dark, the latter reaction being less sensitive. It is obvious, that both *chloranil* and *dichlone* are toxic to plants and especially to algae in the light [235–238]. The molecular mechanism of interaction is, however, not known. Possibly, a spontaneous oxidation of NAD(P)H may occur [234]. Several primary sites of interaction may be involved in the herbicidal effect. Possibly,also a change in membrane fluidity may mediate the herbicidal effect [233].

V. Binding at the "Diuron Site"

1. Native Binding

Binding of photosystem II-inhibiting herbicides to thylacoid membranes may be measured with radiolabeled herbicides or by competition between radiolabeled and unlabeled herbicides. In Fig. 22 the binding of *metribuzin* is compared with the inhibition of photosynthetic electron transport by *metribuzin*, both measurements

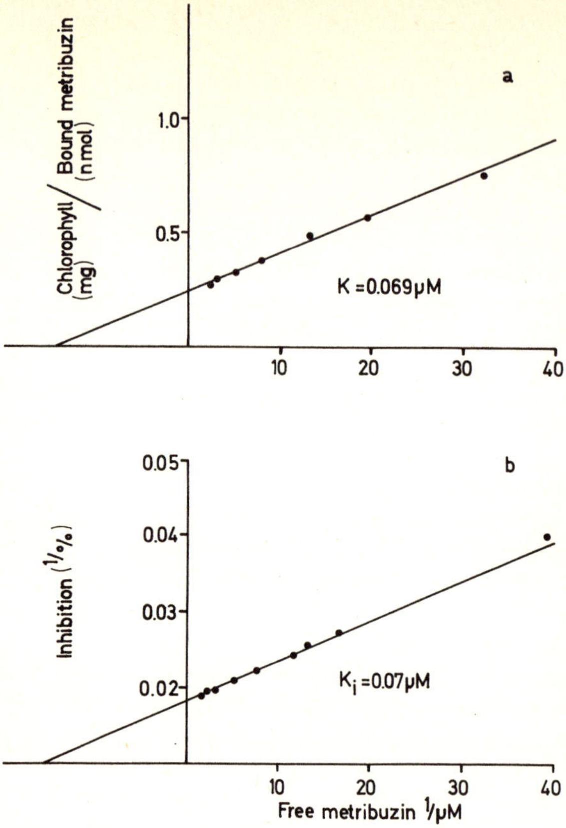

Fig. 22. Double reciprocal plots of *a* binding of labelled *metribuzin* to chloroplast membranes and *b* inhibition of chloroplast electron transport by *metribuzin* [21]

being presented as double reciprocal plots. It is evident that both measurements lead to the same constant for binding and inhibition which means that only specific binding of *metribuzin* to the inhibition site occurs in chloroplast membranes. In other cases, unspecific binding at high concentrations ("low-affinity binding") occurs in addition to specific binding, yielding biphasic binding curves. This is e.g. the case with *atrazine* [21] and *monuron* [89].

Competitive binding experiments have shown that all photosystem II-inhibiting herbicides known so far bind to the same site. However, two groups of compounds may be formed which show competitive binding among compounds belonging to the same group but uncompetitive binding when compounds of the two groups are compared [239, 240, 240 a]. One group is formed by the phenol-type compounds (Tables 12, 13) and the other group by the rest of the respective compounds (Tables 4–11). However, only a few examples have been investigated so far, and other investigators have reported free competition among all the herbicides which inhibit photosystem II [240 c]. As these data show, at least two different proteins participate in the binding, one with a molecular weight of 32,000 dalton

(32 KD protein) and a second with a molecular weight of 41,000 dalton (41 KD protein). As judged from the affinity label, trypsin digestion, and mutant studies, the 32 KD protein might be located on the surface of the thylacoid membrane, whereas the 41 KD protein could be located more in the interior adjacent to the 32 KD protein [240 a] (see below).

As already briefly mentioned, it has been possible to isolate binding proteins by using affinity-labeled herbicides. An azido group has been introduced as an affinity label into a phenol-type herbicide (i-*dinoseb* [239, 240, 240 a]) and also into an *s*-triazine (*atrazine* [241, 242]). After irradiation of thylacoid membranes with either visible or UV-light in the presence of the azido-herbicide, which was radioactively labeled in addition to the azido label, the herbicide was found to have been bound to a "32 KD protein" in the case of azido-*atrazine* and to a "41 KD protein" in the case of azido-i-*dinoseb*. In addition, other proteins had picked up some label which were presumed to be the light harvesting complex (25 KD) and possibly a degradation product (16 KD). The implications of the binding to these proteins will be discussed elsewhere (p. 58).

Maximum binding of the herbicidal inhibitors occurred between 2.4 and 4.0 nmol per mg chlorophyll. Assuming one binding site per photosynthetic pigment complex, this would mean that 300–500 chlorophyll molecules constitute one pigment complex. This value corresponds well with the ratio found for carriers in the electron transport chain of which only one is present per pigment antenna unit and also with the number of chlorophyll molecules per photosynthetic unit obtained by photosynthetic yield measurements.

Theoretical studies of the binding site and mode of action for photosystem II herbicides have been performed by Shipman [243]. He concluded that a hydrophobic component and a flat polar component with a dipole moment of 3–5 D are essential for inhibitory activity. Binding possibly occurs via coulombic interactions of the flat polar component and may lead to a shift of the redox equilibrium at the B-protein (Figs. 14, 15). When speculating about the molecular events occurring during binding and inhibition, bicarbonate interaction must be taken into account. The dependence of photosynthetic electron flow on the presence of bicarbonate has been known for a long time and has recently been reinvestigated [244, 245, 245 a, 302]. The bicarbonate ion was found to compete with *diuron*, *atrazine*, and *DNOC* for the binding site in isolated chloroplasts but not in the blue green alga *Anacystis*. In conclusion, since CO_2 deficiency inhibits photosynthetic electron flow, and since HCO_3^- and herbicides appear to compete for the same binding site, Shipman suggests that an arginine bicarbonate salt may play an important role in stabilizing QB^- during the photosynthetic electron transfer.

2. Herbicide-Insensitive Electron Transport

Several herbicide-insensitive photosynthetic electron transport systems are meanwhile known: (a) after digestion with trypsin, photosystem II becomes directly accessible for the reduction of ferricyanide, (b) treatment with cerulenin appears to alter the lipid composition in some cases, leading to a loss of *diuron* sensitivity, and (c) weed mutants resistant to *s*-triazine herbicides have appeared in the field in many locations. The photosynthetic electron transport system of these mutants is totally insensitive towards *s*-triazines.

a. Trypsin Treatment

After a controlled and partial digestion of thylacoid membranes, the photosynthetic electron transport is no longer sensitive towards *diuron* (Fig. 23). This may be due to the fact that the protein responsible for herbicidal binding and inhibition is located on the thylacoid surface and therefore easily accessible to a proteolytic attack (Figs. 15, 25). As might be expected, this does not only apply to *diuron* but also to other photosystem II herbicides, e.g. *ioxynil* and *metribuzin* [247–249]. However, simultaneously with the early loss of e.g. *diuron* and *atrazine* binding [241], the affinity of the binding site towards phenol-type herbicides is unchanged or even temporarily increased [239, 240 a, 240 b], suggesting an easily accessible binding protein (32 KD) and a second one located deeper in the membrane (41 KD). Both proteins may participate in the binding of and in the inhibition by both herbicidal photosynthesis-inhibiting groups, viz. the *diuron* analogues and the phenols; i.e. more than one protein may constitute the "binding area". These conclusions will later be supplemented by mutant data (p. 55).

When studying the trypsin digestion in the presence and absence of *diuron*, Edelman [250] found distinctive differences in the rate of breakdown and in the pattern of cleavage proteins. The herbicide-binding 32 KD protein, which is one of the protein species digested early by trypsin, is present in the chloroplast in high concentrations. Moreover, it has a very high in vivo turnover rate within only a few hours [251]. It is at present unknown why the protein behaves in this way and what this means for its physiological and/or biochemical functions. Does the high turnover rate actually suggest a possible regulatory function?

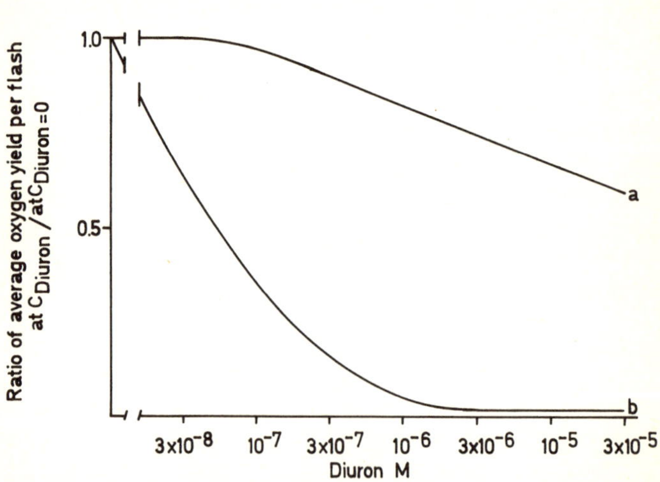

Fig. 23. Insensitivity to *diuron* of oxygen evolution caused by isolated chloroplasts which have been treated with trypsin *a* as compared to untreated chloroplasts *b* [246]

b. Cerulenin Treatment

Cerulenin ((2 S),(3 R)-2,3-epoxy-4-oxo-7,10-dodecadienoyl-amide), an antibiotic produced by the fungus *Cephalosporium caeruleus*, is an inhibitor of fatty acid biosynthesis [252]. In *Chlorella* cells, both oxygen evolution and delayed fluorescence as an indicator of photosystem II activity are no longer affected by 10 μM *diuron* when the cells have been grown for 72 h in 22.5 μM cerulenin [254, 255]. No external electron acceptor was present, and the fatty acid distribution in the thylacoid membrane was changed. This situation reminds of the lipid changes in *s*-triazine-resistant chloroplasts (p. 58) and may be interpreted to indicate an integral function of the lipid components during chloroplast electron transport and its inhibition by herbicides. However, other explanations are possible, and further investigations are needed.

c. Resistance by Mutation

As is well known, the presently available herbicides have been found by screening tens of thousands of chemical candidates, first for herbicidal activity, second for tolerance in a major crop plant. The basis for this tolerance in the case of photosynthesis-inhibiting herbicides is in most cases a difference in the rate of metabolic detoxification, caused either by degradation or by conjugation. Other tolerance mechanisms, e.g. limited uptake and transport of herbicides, have also been reported, but in none of these cases was the sensitivity of photosynthesis altered in the resistant plants. In effect, chloroplasts isolated from resistant crop or weed plants usually proved to be as sensitive as chloroplasts isolated from susceptible plants. Several reviews on the different cases of selectivity and resistance have recently been published [255–257].

As may be expected, new resistant weed varieties can appear when a sufficiently strong selection pressure is maintained for at least several generations, i.e. several years in the case of plants. Gressel and Segel [258] have pointed out the factors which are important for the selection of resistant weeds and calculated the probability curves by computer. The factors studied included: generation time, mutation frequency, selection pressure, fitness of the mutants, soil seed reservoir, and "Parkinsonian plasticity" which measures the adaptability to the environment of the surviving individuals under the conditions of high selection pressure (production of more seeds than normal in this case). Taking these calculations into consideration, the sudden appearence of resistant weed plants in *s*-triazine-treated fields (Table 16) is not surprising. This is all the more to be expected because (a) *simazine* and *atrazine* are fairly resistant to degradation in the soil, thus exerting a long-lasting selection pressure on the vegetation, and (b) because these herbicides have been applied year after year to the same soil, in some cases for up to 12 years. However, this selection occurs at the expense of fitness which is clearly lowered in the resistant plants [259, 259 a].

The mechanism of resistance in *s*-triazine-resistant weeds has extensively been studied. It soon became clear that while uptake, transport, and metabolism of the herbicide were unchanged [260–263], photosynthesis was not inhibited in the resis-

Table 16. Geographical distribution of triazine resistant weeds [255, 255a]

Resistant species	Geographical distribution
Amaranthus retroflexus	U, C, F, S, A
Amaranthus spec. (hybridus, lividus?)	S
Ambrosia artemisiifolia	C, U
Bidens tripartita	A
Brassica campestris	C
Brassica rapa	
Bromus spec.	U
Chenopodium album	U, C, F, S, A, B
Chenopodium spec. (strictum, polyspermum?)	C, F, S
Erigeron canadensis	S
Poa annua	F
Polygonum lapathifolium	F
Polygonum persicaria	F
Senecio squallidus	GB
Senecio vulgaris	U, C, GB
Solanum nigrum	F, B
Stellaria media	B

U, United States; C, Canada; F, France; S, Switzerland; A, Austria; B, Bavaria (southern Germany); GB, Great Britain

Table 17. Resistance factors (I_{50} resistant/I_{50} sensitive) observed in the chloroplasts of *s*-triazine-resistant weeds tested with different groups of photosynthesis-inhibiting herbicides [265, 268, 275]

Inhibitor	Resistance factor
s-Triazines	100–1,000
Pyrazone	90
Metribuzin	30– 300
Bromacil	10– 100
Karsil	9.3
Diuron	1.4
Nitrophenols, bentazone	0.1–0.4

tant plants either in vivo [264] or in vitro [264–269]. It is moreover most interesting that the isolated chloroplasts reflect the differences in sensitivity of the intact plants towards different groups of other photosystem II herbicides (Table 17). The resistance extends to other *s*-triazines, to uracil and *as*-triazinone herbicides, to pyridazinones and to some extent also to the anilides. However, the sensitivity towards *diuron* is nearly unchanged (Fig. 24), and the sensitivity towards *bentazone* and nitrophenols even appears to be increased. This behaviour in part resembles the results obtained from studies on the binding of herbicides to wild-type chloroplasts (p. 51) and to trypsinated chloroplasts (p. 54).

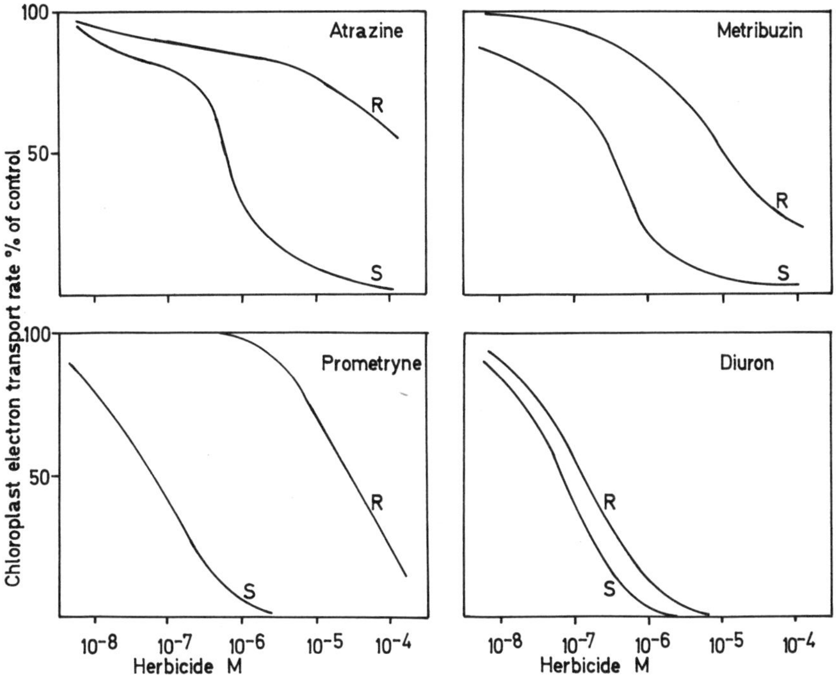

Fig. 24. Inhibition of dichlorophenolindophenol (DCPIP) reduction by chloroplasts isolated from *s*-triazine-sensitive (S) and resistant (R) weed plants by different photosystem II herbicides. *Chenopodium album* chloroplasts were tested with *atrazine* and *metribuzin* [268], and *Senecio vulgaris* chloroplasts were tested with *prometryne* and *diuron* [265]. Reproduced with permission of the Weed Science Society of America

Besides these chloroplast-based resistances to *s*-triazine herbicides, it has been possible to obtain resistant varieties of *Poa annua*, *Echinochloa crus-galli*, and *Setaria viridis* after repeated applications of *metoxuron* [270] and of *Echinochloa crus-galli*, *Setaria viridis*, *Digitaria sanguinalis*, and *Sorghum halepense* after repeated applications of *atrazine* [271]. In contrast to the tolerances observed for triazines quoted in Table 16, where the resistant plants appeared from one year to the next and instantly displayed a high degree of resistance (Table 17), in these cases the observed tolerance appeared gradually and increased over the years. This may be due to a yearly increase in detoxifying capacity, i.e. the selection and propagation of individuals which did not differ in quality but in quantity of the relevant gene expression.

It has also been possible to select *diuron*-resistant strains in the unicellular algae *Chlamydomonas reinhardii* [272], *Bumilleriopsis filiformis* [273], and *Euglena gracilis* [274]. The *diuron*-resistant *C. reinhardii* line is less resistant to the inhibitors of photosynthesis *simazine* and *atrazine* then to *diuron* and *monuron*, but normally sensitive towards *bromacil*. This behaviour is contrary to the *s*-triazine-resistant weeds described above. The *diuron*-resistant *E. gracilis* line is also resistant to o-phenanthrolene; other inhibitors have not yet been tested. In conclusion, several different mutations leading to the loss of herbicide binding and inhibition appear to be possible, each with its own peculiar alteration of the sensitivity spectrum. The selection is particularly simple with unicellular green algae because of the large number of individuals.

3. Models of the Binding Site

The results obtained with trypsin-treated chloroplasts have fairly early led to a model envisaging the binding site as a "protein shield" with a regulatory function in the photosynthetic electron flow [246]. This "protein shield" was thought to contain the herbicide-binding site and to undergo a steric change when binding occurs.

Our present view of the site of action of photosystem II inhibitors has gained considerable detail concerning this "protein shield" [2]. The detailed study of trypsin treated chloroplasts, supplemented by information from affinity labeled herbicides, suggests the participation of two proteins in herbicide binding, a "32 KD protein" in the case of the "classic" photosystem II inhibitors (Tables 4–11) and a "41 KD protein" (or possibly 42 KD) in the case of the phenols (Tables 12, 13) [240a]. However, both herbicide groups affect the same biochemical step in the photosynthetic electron transport system. The questions for future work concern the natural function(s) of these binding proteins. Current evidence suggests a regulatory function for the surface exposed 32 KD protein; the 41 KD protein, which appears to be located in the membrane "below" the 32 KD protein, might well participate in this regulatory function. Possibly then, the herbicides discussed here affect an intricate control system regulating the photosynthetic electron flow rate.

The results obtained with chloroplasts from *s*-triazine-resistant weeds have confirmed and extended this model. Arntzen and coworkers [265, 267, 275, 276] have suggested a binding area which includes a central "essential element domain" obligatory for all inhibitors bound (Fig. 25 II). However, different groups of photosynthesis inhibitors are thought to occupy different overlapping regions of the outer binding area ("specificity domain"). With this model, mutative changes in one part of the binding area would eliminate binding and inhibition by one inhibitor group and affect only slightly or not at all binding and inhibition by other groups. This is exactly what has been observed (Table 17). It may be possible however, as is the case with nitrophenols and *bentazone*, that, after mutation, binding and inhibition by one group of compounds is increased. This model, which should of course be seen three-dimensionally, closely corresponds to the model of Trebst (Fig. 25 I, cf. Fig. 15), which is drawn in only one dimension, and to the "subreceptors" postulated by Draber et al. [277].

The separation of the thylacoid proteins from *s*-triazine-resistant plants on polyacrylamide gels did not reveal any changes in the molecular weights of the proteins [265] but did show subtle changes in the mobility of two native polypeptides [241, 267]. Thorough investigation and characterization of the binding proteins in the near future will certainly provide information on possible changes in the amino acid composition of the 32 KD protein of *s*-triazine-resistant plants and possibly also on the endogenous function(s) of this protein [251].

In addition, membrane lipids may play an important role in herbicide binding. Arntzen et al. [267] concluded from herbicide-membrane preincubations that bind-

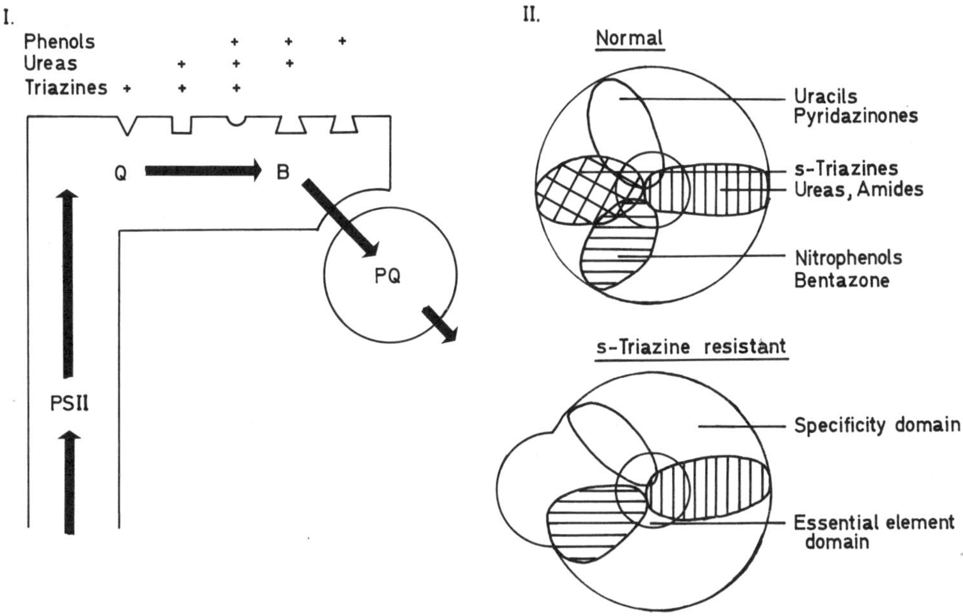

Fig. 25. Models of the binding site which have been designed to explain the different sensitivity changes observed in *s*-triazine-resistant chloroplasts as compared with other photosystem II herbicides. I [1], II [275]

ing is influenced by the physical state of the lipids in the membrane. The lipid composition of the thylacoids of *s*-triazine-resistant plants is also significantly changed [278]. The results obtained with algae even more clearly demonstrate the involvement of membrane lipids in herbicide binding. Thus, experiments with cerulenin (p. 55) and studies of *diuron*-resistant *E. gracilis* confirm the importance of the native lipid composition in the thylacoid membrane for inhibition to occur. In *E. gracilis* adapted to 25 μM *diuron*, no protein changes or protein involvement in the resistance to herbicidal action could be demonstrated. On the contrary, the thylacoid lipids differed in amount and composition leading to more fluid membranes in the resistant line [274].

In conclusion, the binding site for photosystem II herbicides is located on the thylacoid surface and comprises at least two proteins with 32 and 41 kD molecular weights with the former on the membrane surface. In the scheme represented in Fig. 15 this is the region of the "B-protein". Binding is thought to lead to a change in the redox potential or to an equilibrium shift of the plastoquinone B [25, 243] which is then no longer able to participate in the photosynthetic electron transport. The molecular surroundings of the binding proteins are presumed to include a specific arrangement of membrane lipids. The function of the lipids could be that of a spacer or specific "glue" which controls the interaction between the participating polypeptides and the electron carrier molecules. The natural function of the herbicide-binding site could be that of a regulatory binding site for CO_2 (p. 53).

A second binding site for photosynthetic inhibitors apparently exists on the other, i.e. reducing side of the plastoquinone pool (Fig. 15). Inhibitors binding at this second site do not compete with the binding of the herbicidal photosystem II inhibitors and are no commercial herbicides. They include the plastoquinone antagonist DBMIB (p. 24) and some "dinitrophenyl ethers" [29, 247].

VI. Light Excitation of Photosynthetic Pigments and Photosynthetic Oxygen Reduction

Under certain conditions, oxygen may only be partially reduced and then gives rise to extremely reactive and potentially dangerous molecular species which pose a threat not only to anaerobic but also to aerobic organisms. However, whereas anaerobes suffer severe damage from oxygen, aerobic organisms survive in our normal oxygenous atmosphere because they have evolved several protective mechanisms. Still, plants kept in an atmosphere containing much more than the usual 20% oxygen suffer damage and will eventually die.

During the process of photosynthesis, oxygen is not only liberated by the water-splitting enzyme system Y (Fig. 15) but may also be partially reduced by the excited photosynthetic pigment system, yielding highly reactive molecular species. After this knowledge became established in recent years, it also became increasingly clear that photosynthetically activated or partially reduced oxygen species play a key role in the generation of phytotoxicity after the application of herbicides which interfere with either the photosynthetic electron transport (PS II inhibitors, electron acceptors at the PS I) or with the photosynthetic pigment system (inhibitors of carotenoid biosynthesis, diphenylethers). A more detailed description of the process of photosynthetic oxygen "activation" is therefore given in the following.

When a chlorophyll molecule absorbs a light quantum (photon), one of its electrons adopts a higher energy level. The resulting excited singlet state (^{1}Chl) is very labile, and the electron may fall back into the ground state and emit the energy difference as fluorescence or thermal energy in 10^{-9} s (Fig. 26). The emitted wavelength will be somewhat higher than the absorbed wavelength because the electron will lose some energy to molecular rotations, oscillations, etc. However, one electron in the singlet-state chlorophyll may also change its directional spin, thereby yielding a triplet chlorophyll (^{3}Chl). The longer lifetime of the triplet chlorophyll ($\geqq 10^{-4}$ s) greatly increases its chance to perform chemical work – by transferring its excitation energy/an electron to another molecule. However, the close association of excited chlorophyll with other molecules in the thylakoid membrane ("charge-transfer complexes") may also cause the singlet state to do chemical work.

Basically, the chemical work performed by an excited chlorophyll molecule is an electron transfer. The excited chlorophyll molecule loses a high-energy electron to an acceptor A which is reduced to A$^{\cdot}$. The remaining Chl$^+$ is a very strong oxidant and replaces the lost electron from a donor D$^{\cdot}$ which is oxidized to D. As a result, the light energy originally contained in the absorbed photon (h × ν) is partially conserved in the molecule A$^{\cdot}$ (AH after absorbing a proton):

$$h \times \nu + A + D^{\cdot} \rightarrow A^{\cdot} + D + \text{heat}.$$

Under in vivo conditions, D· will be water and A will be Q for photosystem II and plastocyanin and X (or ferredoxin) for photosystem I, respectively (Fig. 14).

A fundamental problem arises for the green plant from the simultaneous occurrence in close proximity (in the chloroplast thylakoid) of highly reactive chlorophyll states and molecular oxygen. All four oxygen species which may be formed (singlet oxygen, $^1O_2$; superoxide anion, $O_2^{\cdot -}$; hydrogen peroxide, H_2O_2; hydroxyl radical, $OH^{\cdot}$) are potentially very harmful. In the case of accumulation in the tissue, they lead to pigment breakdown (Fig. 33, p. 92), lipid peroxidation (p. 68), and eventually to membrane and tissue disintegration. However, the plant has evolved several protective systems which effectively either scavenge or destroy the harmful oxygen species mentioned above, or, in the case of $OH^{\cdot}$, avoid its generation:

(a) β-carotene and α-tocopherol which react with $^1O_2$ [3],
(b) superoxide dismutase (SOD) which produces oxygen and H_2O_2 from $O_2^{\cdot -}$ [4],
(c) catalase which converts H_2O_2 to water and oxygen, and
(d) peroxidases which reduce H_2O_2 to water, using a variety of reductants available in the cell [6]:

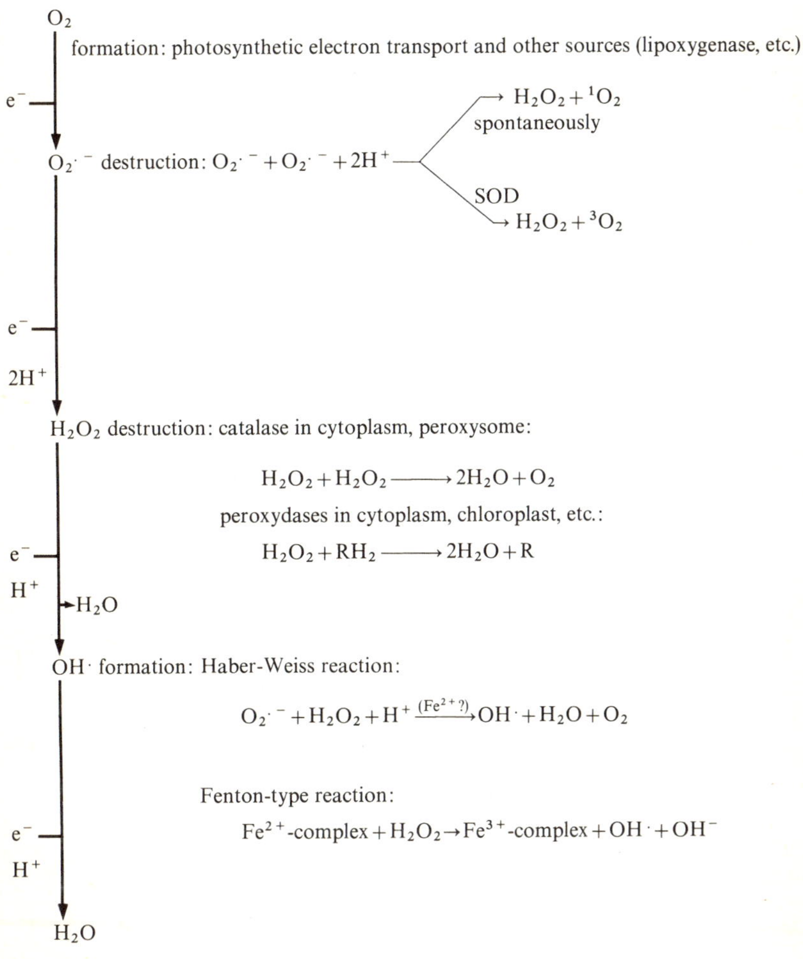

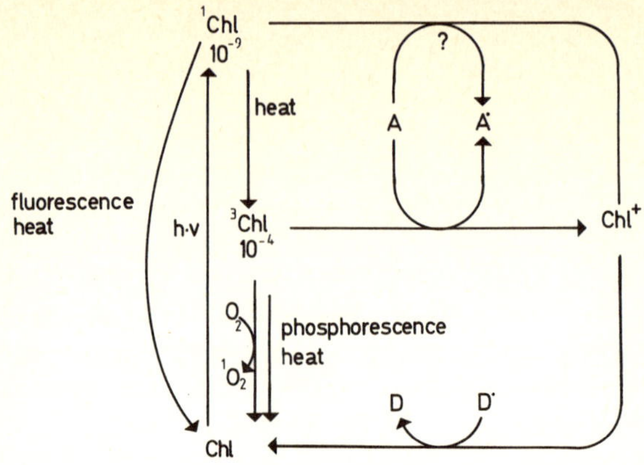

$$\sum h \cdot v + A + D' \longrightarrow A' + D + \text{heat}$$

Fig. 26. Excitation of chlorophyll by absorption of photons ($h \cdot v$) and subsequent possible reactions and transfers (see text)

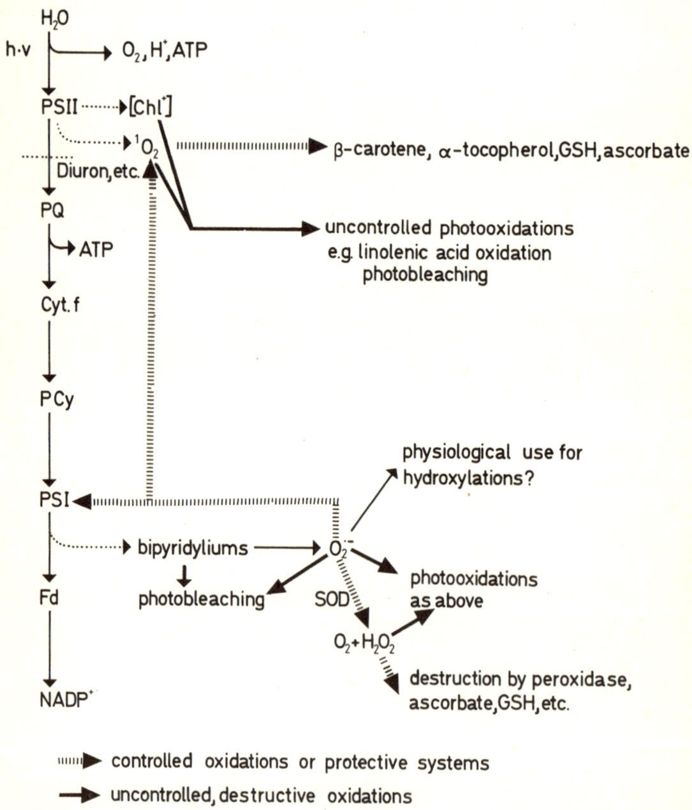

Fig. 27. Generation and degradation of reactive oxygen species in the chloroplast including controlled (protective) and uncontrolled (destructive) reactions. SOD, superoxide dismutase; GSH, reduced glutathione

An OH· radical-type may be formed by a one-electron transfer onto H_2O_2. Of the two reactions described in the above scheme, the Haber-Weiss reaction is highly unlikely to occur in vivo because of the extremely low $O_2^{\cdot -}$-level in the presence of superoxide dismutase and because of a low rate constant [5]. Moreover, the Haber-Weiss reaction may not occur at all in a non-catalyzed manner. The Fenton-type reaction may proceed with low molecular weight complexes of Fe^{2+} or with reduced ferredoxin [5a]. However, the availability of the Fe^{2+}-complexes in vivo in the intact leaf is not clear. The Fenton-type reaction may be obtained in vitro with e.g. the following electron donors: H_2O_2, Fe^{2+}-EDTA, ascorbate, ferredoxin, semiquinones [5a, 7].

The formation and destruction of reduced oxygen species relevant for the present discussion is summarized in the scheme on p. 61 and in Fig. 27. The formation of $O_2^{\cdot -}$ from molecular oxygen requires a univalent reduction, i.e. a one-electron transfer. Besides, inter alia, lipoxygenase and peroxidases [5, 16], the reduced pigment system I/reduced ferredoxin is the most important system producing $O_2^{\cdot -}$ in vivo. Univalent oxygen reduction and production of $O_2^{\cdot -}$ appears to occur under in situ conditions when the NADPH/NADP ratio exceeds ca. 0.8. The resulting "pseudocyclic electron flow" (Fig. 15), where ATP-formation still takes place but no net oxygen change is measured, has been termed "flexible coupling" [6, 8] (i.e., "pseudocyclic electron flow" is an open electron transport from water to an autoxidizable electron acceptor which fails to produce a net oxygen change; during "pseudocyclic electron transport" in isolated chloroplasts the H_2O_2 produced by the autoxidation process is converted into H_2O and O_2 by the contaminant catalase, thus balancing the O_2 produced during the water splitting process). A situation where this occurs in the normal plant life is during the beginning of the light phase [9]. The plant must therefore be able to cope with low ("normal") rates of production of $O_2^{\cdot -}$ and also of H_2O_2.

In contrast to the photosystem I, which catalyzes an univalent oxygen reduction, photosystem II appears to catalyze a divalent oxygen reduction to yield H_2O_2. This, however, is only observed with certain artificial electron acceptors in vitro [10].

To sum up, the following protective processes occur in green plants:

1. Quenching of singlet oxygen by β-carotene, α-tocopherol, ascorbic acid, reduced glutathione, etc. [3, 11–14]. Singlet oxygen may either be produced by energy transfer from triplet chlorophyll (Fig. 26) or from $O_2^{\cdot -}$ after feeding an electron into one of the pigment systems [3–5]. When $O_2^{\cdot -}$ is generated at excessive rates, "photo-bleaching" occurs. β-carotene may also directly quench chlorophyll in its triplet state.
2. Enzyme superoxide dismutase. Besides effectively removing $O_2^{\cdot -}$, enzymatic dismutation may have the advantage of producing triplet instead of singlet oxygen [3–5]. It has been calculated that the spontaneous lifetime of $O_2^{\cdot -}$ would allow a 5 μm diffusion distance to exist, which would mean far-reaching cellular destructions [15]. However, superoxide dismutase, which is present in most aerobic organisms and in all higher plants, increases the spontaneous rate of $O_2^{\cdot -}$ dismutation 10^9-fold at an anticipated steady-state level of $O_2^{\cdot -}$ of 0.1 nM [16].

3. Destruction of hydrogen peroxide by catalase which, however, is not present in the chloroplast [6]. Only low H_2O_2 concentrations may therefore be tolerated by the chloroplast [5b]. However, the peroxisome as a primary catalase site is commonly very closely associated with the chloroplast membrane.
4. Cooperation of ascorbate and glutathione in the intrachloroplastic reduction of H_2O_2 via peroxidase [17]. The dehydroascorbate formed in this reaction is re-reduced by glutathione which, in turn, is possibly re-reduced by NADPH [6].
5. A "xanthophyll cycle" whose function in vivo is, however, not clear [6, 18]. The xanthophyll cycle incorporates oxygen into the epoxy groups of antheraxanthin and violaxanthin. The de-epoxidation and resulting formation of zeaxanthin is a light dependent reaction. The cycle occurs both in the thylacoid and in the chloroplast envelope [18, 18a]. However, the rate of oxygen uptake attributable to this cycle is rather low [18b].
6. Repair reactions which may play a role and should be considered when summarizing the defense arsenal of plants against lethal photooxidations (p. 65). Photorespiration (p. 20) may also contribute significantly to the destruction of excess reducing power.

VII. Toxicity after Inhibition of Photosystem II

1. General

Only a few years after the detection of their herbicidal activity, urea herbicides [279, 280] and *s*-triazine herbicides [142] have been found to exhibit photosynthesis-inhibiting activity. However, a possible cause for the damage observed in the intact plant a few days after treatment with a photosynthesis-inhibiting herbicide has been found only recently.

Generally, yellowing of the leaf veins and necrosis at the leaf margins and in the intercostal areas is observed [171]. This pattern may be explained by the distribution of the herbicide in the leaf after application to the soil: the herbicide is transported in the leaf along the veins and tends to be concentrated at the margins. When the herbicide acts slowly and the supply of the leaf with herbicide is rather low, then effects become gradually visible only along the leaf veins. When the herbicide acts rapidly and its concentration in the leaf is constant and high, then toxicity is observed in the intercostal areas where the rates of photosynthesis are highest. Slow action may produce light green to yellow or even white tissue indicating chlorophyll destruction, whereas rapid herbicide action is accompanied by wilting and necrosis and leaves little time for total chlorophyll destruction. Bleaching may frequently be observed with some photosystem II inhibitors and only occasionally with other inhibitors. Consequently, the visual damage provides little information concerning the underlying molecular events.

Early experiments have already revealed that the lack of assimilates in the inhibited plants does not cause the observed damage; i.e. the plant does not die of starvation. One point was that herbicidally treated leaves supplied with sucrose [142] or with other sugars [31] did form starch, indicating intact metabolic pathways besides the inhibited light reaction; furthermore, the supply of sucrose with herbicidally treated plants did not avoid the death of the plant but only delayed it [59]. Also, plants treated with *diuron* and kept in the dark for up to 7 days did not show any damage when treated with the herbicide whereas plants treated similarly but exposed to full day-light wilted and died after 3–5 days.

Clearly, then, starvation is not a factor in the herbicidal action of photosynthesis-inhibiting compounds. This has also been concluded very early by Sweetser et al. [281], who postulated a phytotoxic substance which is produced when photosystem II is inhibited in the light.

2. Physiological Conditions Influencing the Appearence of Damage

Light is of course the most prominent factor influencing the degree of damage. Numerous investigations have demonstrated that the more light is applied, the more extensively and the more rapidly damage will occur [151, 155, 156, 282]. A consequence of this dependence may be seen in the rates of the application of these herbicides which are considerably lower in southern countries where high light intensities prevail. More specifically, the highest damage will occur when the light intensity was low before and high after the application of the herbicide (Table 18) [151]. Two lines of evidence are important for the understanding of these results: (a) every plant exposed to low intensity light will form a "low-light plant" or "shade plant" (p. 71); this plant is of course designed to handle low light intensities more efficiently than the normal "sun-type plant"; (b) when a "shade-type plant" is exposed to high light intensities in a photosynthetically inhibited state, its pigment system is even more overloaded with excitation energy than in a "sun-type plant", i.e. it reacts more sensitive.

Of course, also temperature and humidity influence phytotoxicity. Whereas toxicity increases with rising temperature [155, 156], it decreases with growing humidity. The latter is simply the consequence of a decreased flow of water through the plant when the air is humid. Conversely, unusually high amounts of herbicide may be taken up in warm and dry air as long as the water supply of the roots is not limited, resulting in phytotoxicity at lower than usual application rates. This is possible because most of the photosystem II inhibitors move freely with the transpiration stream in the plant.

Table 18. Dry weight development and ratings of damage for *metribuzin*-treated tomato plants kept under different light conditions [169]

Light intensity (lx)		Dry weight		Rating of damage
Before treatment	After treatment	g	%	%
22,000	22,000	1.97	56.3	36.2
6,500	6,500	1.12	66.2	50.2
22,000	6,500	1.66	70.0	15.4
6,500	22,000	1.19	36.8	68.3

It has occasionally been argued that plants fixing CO_2 via the C4-pathway are inherently less sensitive towards photosystem II inhibitors than C-3 plants [91, 119, 283]. The basic idea is that the process of photorespiration, which only occurs in C3-plants, rapidly depletes the freshly fixed carbon skeletons via the glycolic acid pathway, initiating the early appearence of phytotoxicity because of the resulting more rapid loss of carbon compounds as compared to C4-plants. The effect of sucrose supplied together with the application of a photosynthesis-inhibiting herbicide may be interpreted similarly: as long as the inhibited plants contain freely available and respirable carbohydrates they may support repair reactions which delay and/or decrease the tissue damage. In summary, the appearance of phytotoxicity may be modulated by factors such as the physiological status of the pigment system, the distribution of excitation energy among different pathways (p. 63), and the carbohydrate status of the cells.

Van Oorschot [150] has shown that essentially similar damage symptoms are observed when a plant is either treated with a photosynthesis-inhibiting herbicide or kept in CO_2-free air. These data demonstrate that, when the excitation energy generated in the photosynthetic pigment complexes cannot be transferred to the native acceptors and finally to CO_2, other harmful pathways of energy dissipation will occur. These will be dealt with later (p. 68).

3. Ultrastructural Damage

The ultrastructural detail which first shows damage in an *atrazine*-treated plant is the intergranal thylacoid membrane [284, 285]. With progressing deterioration the whole chloroplast gains a swollen appearance, and the granal structure becomes disintegrated [286]. The other subcellular structures, including mitochondria and tonoplast, eventually become disintegrated as well. It should, however, be stressed, that this type of damage only occurs during rapid herbicidal action. Sublethal herbicidal concentrations not destroying but only stressing the tissue may lead to adaptive ultrastructural changes (p. 71).

4. Possible Toxicants

a. Energy Availability

The fact that damage from photosystem II-inhibiting herbicides can at least be delayed by the application of sucrose or other "respirable carbohydrates" has been interpreted in favour of a predominant role of the energy state in the preservation of an intact cell [288]. True as this is, it cannot easily explain the cause of the actual damage, but only why damage does not occur under certain circumstances (p. 65).

b. Nitrite

The reduction of nitrate in plants proceeds in two steps: nitrate reduction by cytoplasmic nitrate reductase forming nitrite, and nitrite reduction by chloroplast ni-

Table 18a. Formation of nitrite in *Sinapis alba* leaves in the light under air and nitrogen after treatment with different herbicides [291]

Herbicide	pI_{50}	NO_2^- (nmol per g fresh weight)	
		Air	N_2
Chloramben	<4	0	0
2,4-D	<4	0	40
Dinoseb	5.0	408	72
Diuron	6.4	13	600
Ioxynil	6.0	224	64
Lenacil	6.6	0	600
Methabenzthiazuron	6.4	8	600
Metribuzin	6.6	10	600
PCP	5.2	18	124
Simazine	5.7	20	600

trite reductase forming ammonia. Since the chloroplast nitrite reduction depends on photosynthetically reduced ferredoxin, the inhibition of photosynthesis by herbicides may block nitrite reduction and lead to the accumulation of nitrite in the cell. Nitrite is known to be a toxic substance, and Klepper therefore suggested that the herbicidal damage of the plant observed after inhibition of photosynthesis may be caused by nitrite [288–290].

Whereas the inhibition of nitrite reduction by photosynthesis inhibitors has been demonstrated both in vitro and in vivo, the accumulation of nitrite in aerial leaf tissue occurs only rarely [292]. Three conditions favour the production and accumulation of nitrite in photosynthetically inhibited leaf tissue: (a) simultaneous stimulation of NADH availability for nitrate reduction by respiratory uncouplers [289], (b) irradiation in a nitrogen atmosphere (Table 18a), and (c) irradiation in the presence of high contents of nitrate [293]. Nitrite may also accumulate in leaf tissue in the dark when the plant is kept in a nitrogen atmosphere [294]. However, it has also been concluded from experiments performed under nitrogen that the nitrite contained in the tissue is not responsible for the observed damage [291]. Experimental conditions under which accumulation of high levels of nitrite occurs in the leaves have been found without causing any damage to the latter. In a study with *ametryne*, Churchill et al. [290] have found increased damage in plants grown on nitrate vs. ammonia or without nitrogen. Of course, only the plants grown on nitrate accumulated nitrite. However, the plants grown on ammonia were also subject to damage, albeit more slowly.

Interestingly, not only photosynthesis-inhibiting herbicides but also uncouplers of oxidative phosphorylation (*dinoseb* and *ioxynil*, Table 18a) induce the accumulation of nitrite in leaves, but preferably in the air. This could be due to the uncoupling effect which may lead to higher concentrations of NADH and higher in vivo nitrate reductase activity.

c. Radicals and Activated Oxygen Species

Several authors have found that in isolated chloroplasts as well as in intact leaves β-carotene is destroyed prior to chlorophyll (Fig. 29) [295–297]. Simultaneously, malondialdehyde (MDH), and ethane are formed as degradation products of the membrane lipids [10, 298, 299]. This degradation is thought to proceed via lipid peroxidation after singlet oxygen ($^1O_2$) or radical (e.g. Chl^+) attack at the lipid (cf. p. 60):

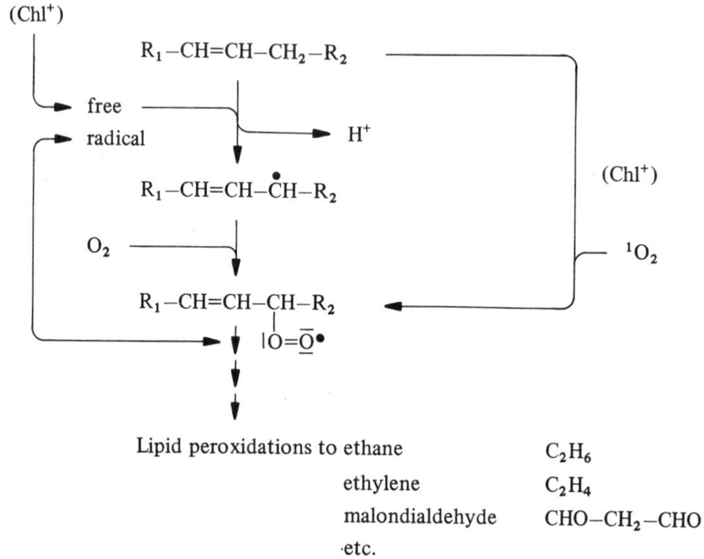

The involvement of $^1O_2$ in the observed oxidative reactions may be concluded from the observation that DABCO (1,4-diazabicyclo-(2,2,2)octane), a powerful quencher of singlet oxygen, prevents the formation of ethane and other destructive effects [296, 300]. $^1O_2$ may e.g. be formed by energy transfer from triplet chlorophyll (Figs. 26, 27, 28). However, the excited chlorophyll molecule (Chl*, or Chl^+ after losing an electron) may also directly initiate a chain reaction by abstracting an electron from unsaturated fatty acids. The radical chain reaction thus started would lead to a similar lipid peroxidation.

The sequence of events involved in the generation of primary phytotoxicity from photosystem II-inhibiting herbicides could then tentatively be formulated as follows (Figs. 28, 29): the excited chlorophyll pigment normally transfers the absorbed energy to the associated electron transport chain. If this, however, is not possible, e.g. due to herbicidal inhibition or lack of electron acceptors, the excitation energy may be absorbed by several natural pathways functioning as "safety valves". One possibility is the emission of excess excitation energy as fluorescence. In effect, fluorescence is most intense when photosynthesis is inhibited (Fig. 20). The transfer of excitation energy to oxygen with the formation of singlet oxygen is potentially very dangerous because of the high reactivity of $^1O_2$. Naturally occurring $^1O_2$ is very effectively quenched by e.g. β-carotene (p. 63). However, excessive amounts may not be removed sufficiently rapidly and may eventually oxidize

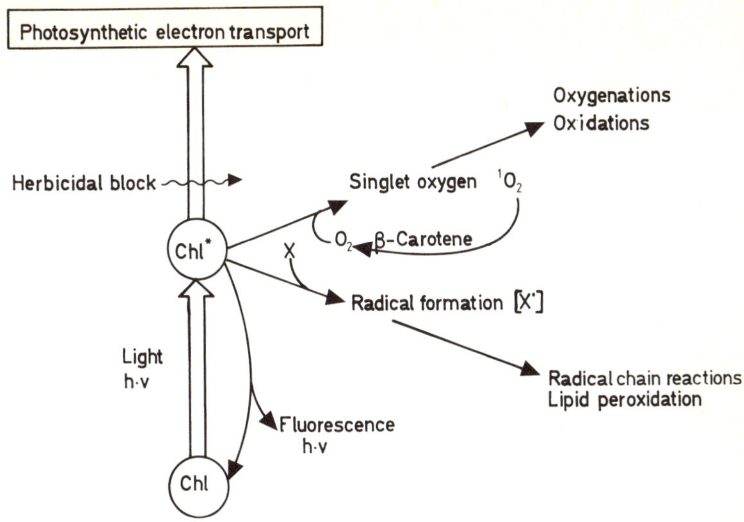

Fig. 28. Excitation energy spill-over from excited chlorophyll (Chl*) under conditions of herbicidally blocked photosynthetic electron transport

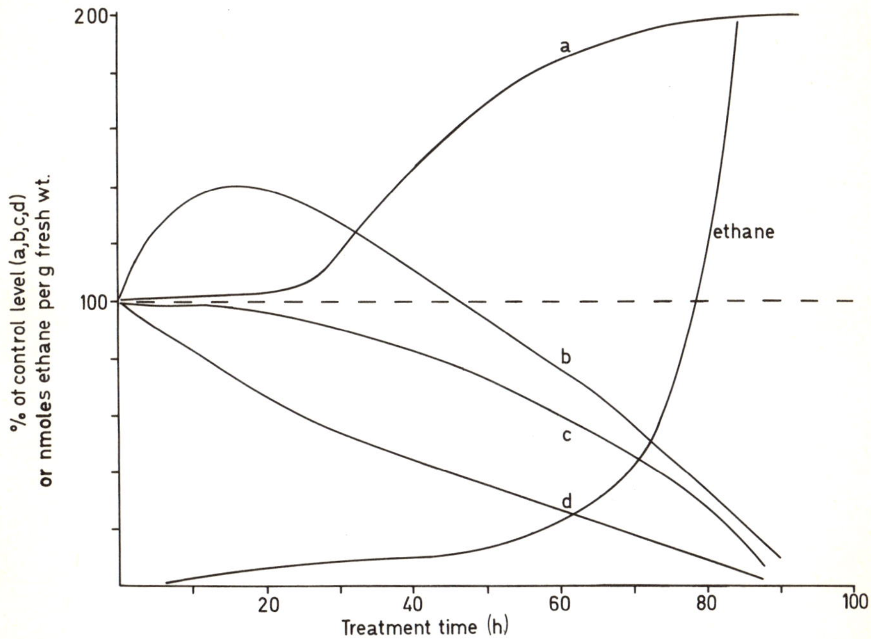

Fig. 29. Changes of malondialdehyde production *a* and of levels of lutein *b*, chlorophyll *c*, and *β*-carotene *d* as well as ethane production in *monuron*-treated flax cotyledons [296, 301]

fatty acids, pigments and other membrane components. Also, a more direct radical formation via Chl^+ may occur and initiate the oxidative chain reaction described above. This reaction is autocatalytic because the oxygenated fatty acids can apparently abstract electrons from other unsaturated fatty acids.

VIII. Physiological Effects Induced by Herbicides that Inhibit Photosynthesis

1. General

Among the photosynthesis-inhibiting herbicides, *s*-triazines became early known for their "greening effect" and for their growth-stimulating activity [304]. The striking observation that *simazine*- and *atrazine*-treated plants grew higher and looked darker green has aroused the interest of farmers and researchers alike. Also, the senescence of early leaves and of decaying plants at the end of the growing season was delayed. However, similar observations have also been made with urea herbicides, *as*-triazinones, and other photosynthesis-inhibiting herbicides as well [186].

These so-called "physiological effects" are frequently observed in the crop plant, but any other plant surviving the herbicide action may respond similarly. Indeed, physiological effects could also be termed "sublethal effects" because the plant is usually not visibly damaged but instead shows morphological, physiological or biochemical responses after herbicide application.

Two molecular mechanisms have been suggested for the interpretation of the physiological effects:
(a) the effects may be regarded as regulatory responses caused by a low photosynthetic rate and a decreased carbohydrate production in the herbicide-treated plants or (b) they may be the consequence of a hormone-like activity of the herbicide molecule itself. Several observations favour the interpretation as a regulatory phenomenon:

(aa) the non-photosynthesis-inhibiting hydroxytriazine analogs show no effects [305, 306];
(ab) all photosynthesis-inhibiting herbicides from structurally different groups tested so far induce the "greening effect" and the "shade adaptation";
(ac) in all cases studied a decrease of carbohydrate content has been found to accompany the "greening effect";
(ad) in order to induce the effect with seedlings from large seeded plants, the reserves of the seedlings have to be eliminated shortly after germination of corn [307] or soybean [308] plants,thereby enhancing the decrease in carbohydrate content caused by the herbicide,
(ae) *s*-triazine-resistant weeds do not accumulate nitrate after treatment with atrazine [309].

(b) However, a number of arguments, which only apply to *s*-triazine herbicides, suggest that hormone-like effects are involved in the observed growth stimulation:

(ba) *s*-triazine herbicides exhibit cytokinin activity in a number of assay systems (Table 21),
(bb) *atrazine* stimulates growth and other parameters in non-photosynthetic tissues, possibly by exhibiting cytokinin activity.

Some of the observed physiological effects may be interpreted either as a regulatory response or as a hormone effect, e.g. the delayed senescence of leaves after treatment with *methabenzthiazuron* [310] or with *atrazine*, *bromacil*, and *fluometuron* [186] or the increase of the concentration of soluble proteins after treatment with *methabenzthiazuron* [311] or *simazine* [312]. However, both effects are probably regulatory responses since they occur with different classes of photosynthesis-inhibiting herbicides. Still, the chain of effects (p. 11) may include a change of the endogenous plant hormone levels [313].

Other physiological effects are more clearly regulatory responses, like e.g. the formation of shade-type chloroplasts in plants treated with photosynthesis-inhibiting herbicides (Table 19). Still other physiological effects, like e.g. the increase of the rate of protein synthesis [314] and of the rate of RNA synthesis [131] found in etiolated tissue after *atrazine* treatment may be conveniently explained by a hormone-like activity of the *s*-triazine molecule itself.

2. Regulatory Responses after Partial Inhibition of Photosynthesis

The most important characteristic of regulatory responses is the considerable time delay between herbicidal application and the eventual effect. This characteristic distinguishes them from the direct effects of herbicidal compounds. Therefore, a central problem arising during the study of regulatory responses is the standardization of the experimental conditions. The best approach for obtaining reproducible physiological data is to control all environmental parameters during the course of the experiment. The duration of these experiments may range between one day and one or two weeks and the sequence of primary, secondary and tertiary effects can be determined by kinetic analysis (Fig. 5). Field experiments, on the other hand, suffer from widely differing environmental conditions and from the resulting low reproducibility. Therefore, the results obtained and conclusions drawn from controlled physiological experiments should be considered before the interpretation of field data.

Chloroplasts

The first effect observed in vivo after the application of a photosynthesis-inhibiting herbicide is, as might be expected, the inhibition of photosynthetic CO_2-fixation resulting in a decrease in the concentration of soluble sugars (Fig. 30). Subsequent changes are observed after a few days up to one week in the chloroplast anatomy and morphology (Table 19) and in the nitrogen metabolism (Table 20) [318]. Most significantly, the chloroplast morphological, anatomical, and biochemical changes induced by either light of low intensity or by herbicidal treatment are identical. This interesting observation has led to the suggestion that the herbicide-treated plants may in effect resemble "shade-adapted" or "low-light adapted" plants [310, 318]. A growing number of herbicides, only having in common photosynthetic electron transport as the herbicidal action, behave similarly in this respect (Tables 19, 20).

The "shade-type chloroplast" and the response of plants to either white light of low intensity or to natural shade has been the subject of recent reviews [322, 327, 350, 351]. In the shade-type chloroplast, up to three light harvesting pigment complexes are formed per electron transport chain (Fig. 14), thereby ensuring the maximum possible conversion of light energy into chemical energy. The consequence is an increase of the chlorophyll content but a decrease in some carriers of the elec-

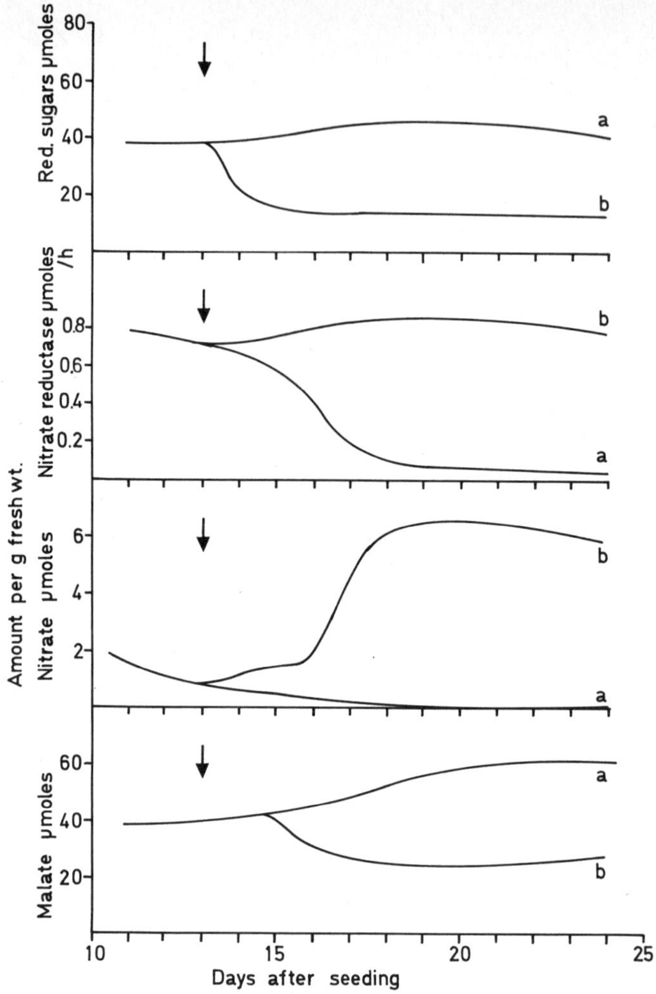

Fig. 30. Kinetic responses of soluble reducing sugars, in vitro nitrate reductase activity, nitrate, and malate in the leaves of wheat plants which were treated with 3 $mg\,l^{-1}$ *methabenzthiazuron* via the roots 13 days after seeding [315] (arrow). *a* control, *b* treated

tron transport chain, namely plastoquinones, cytochromes f and b and ferredoxin (Table 19). Only part of the relevant changes have been observed so far after herbicide treatment, but the results obtained generally agree with those obtained after treatment with light of low intensity or in the shade. Only in the case of *simazine* have some aberrations been observed which may, however, be related to the hormone-type activity of this compound which appears to be independent of the shade response [329].

An interesting case of shade-type behaviour has been reported by Tanaka et al. [352]; they found that *Spirogyra* filaments treated with *diuron* behaved, concerning their mobility, in the same way as if exposed to low-intensity light. This, however, may also reflect a direct control of mobility by the carbohydrate supply.

Table 19. Changes in chloroplast morphology and biochemical composition after treatment with photosynthesis-inhibiting herbicides

Shade-type response	Reproduced with herbicides
Swollen after isolation	Atrazine, methabenzthiazuron
Decrease of stroma lamellae	Bentazone, fluometuron, methabenzthiazuron
Increase of grana stacking	Bentazone, diuron, monuron
Increase of grana/stroma lamellae ratio	Methabenzthiazuron
Increase of grana size	Fluometuron, monuron
Increase of chlorophyll (a + b) per fresh wt. and per chloroplast	Atrazine, diuron, methabenzthiazuron
Decrease of chlorophyll a/b	Atrazine, bentazone, diuron, methabenzthiazuron, metribuzin, simazine
Increase of $\frac{a+b}{c}$	Methabenzthiazuron
Increase of $\frac{x}{c}$	Bentazone, methabenzthiazuron, simazine
Decrease of plastoquinone etc.	Bentazone, methabenzthiazuron, simazine
Decrease of ratio per chlorophyll of cytochromes b, f and of ferredoxin	not shown
Decrease of carboxydismutase	Atrazine, methabenzthiazuron, prometryne, simazine

For references see *atrazine* [317, 319, 320], *bentazone* [313, 321, 322], *diuron* [313, 323, 324], *fluometuron* [73], *methabenzthiazuron* [310, 311, 315, 316, 325, 326], *metribuzin* [327], *monuron* [328], *simazine* [329]
a, b = chlorophylls a and b; x = xanthophyll; c = β-carotene

Table 20. Changes in nitrogen metabolism in plants treated with photosynthesis-inhibiting herbicides

Response at low light intensity	Reproduced with herbicides
Malate conc. decreased	Atrazine, methabenzthiazuron, metribuzin, phenmedipham
Nitrate uptake increased	Atrazine, linuron, simazine
Nitrate conc. increased	Atrazine, diuron, lenacil, linuron, methabenzthiazuron, metribuzin, pyrazon, simazine
Nitrate reductase activity increased	Atrazine, diuron, methabenzthiazuron, metribuzin
Soluble amino acid content increased	Methabenzthiazuron, monolinuron, several *s*-triazines
Soluble protein content increased or decreased after prolonged exposure to low-intensity light	Lenacil, methabenzthiazuron, *s*-triazines

For references see *s-triazines* [186, 305–307, 312, 317, 320, 330–346], *diuron* [52, 347], *lenacil* [348, 349], *methabenzthiazuron* [310, 311, 315, 316], *metribuzin* [316, 327], *pyrazon* [348, 349]

N-Metabolism

A second important area of metabolism affected similarly by either light of low intensity or by treatment with sublethal concentrations of photosynthesis-inhibiting herbicides is the nitrogen metabolism (Table 20). A very prominent effect is the opposite response of the nitrate and malate levels [352] (Fig. 30). According to the work of Hageman's group [355, 356], the endogenous concentrations of these two metabolic compounds are causally related with one another. The increased in vitro nitrate reductase activity presumably results from the increased nitrate levels in the treated plants. However, the in vivo nitrate reductase activity is not always increased but may even be decreased [310, 356] since the NADH required for nitrate reduction is formed by either carbohydrate [357] or malate [355] dehydrogenation, both of which are decreased in herbicide-treated plants. Nevertheless, a strong increase of soluble amino acid, soluble protein, and total nitrogen content has often been reported [353]. Insofar as this increase may be seen as a consequence of an increased supply of reduced nitrogen, it may also be classified as a "shade response". However, many cases have been reported where *s*-triazines led to increased rates of protein synthesis and increased protein content in etiolated tissue (p. 76).

Growth and Yield

An increased protein content in *s*-triazine-treated plants has also been reported under field conditions [358–366]. However, when coupled with yield measurements, the general picture that emerged was that of increased protein content and decreased yield going together, indicating that the protein and nitrogen increases were accompanied by a decreased productivity, presumably caused by partially inhibited photosynthetic CO_2-fixation. In cases of increased yield, the protein content of the plants on a dry weight basis was, on the contrary, unchanged [367]. This correlation led to the classification of the increases of the protein content as a "stress response" – physiologically in this case a "low-intensity light response" – or "hormesis effect". "Hormesis", meaning stimulation by sublethal concentrations of toxic compounds, is obviously a very common effect [304, 368]. In contrast to effects of growth hormones, "hormesis effects" only occur within a narrow concentration range, with increasing concentrations becoming rapidly phytotoxic and/or strongly decreasing the measured parameter below control values (Figs. 9, 31, 32].

$^{14}CO_2$-fixation experiments performed with plants adapted to low rates of photosynthesis-inhibiting herbicides have repeatedly shown an increased flow of fixed carbon products into the soluble amino acid pools [369]. Other authors reported either a dark $^{14}CO_2$-fixation pattern after *atrazine* treatment [370] or simultaneous increases in organic acid labelling and decreases in sugar and amino acid labelling after *prometryne* treatment [371] for shorter herbicidal pretreatment times. Also, a positive correlation between the amount of rough endoplasmic reticulum, which is the location of protein synthesis, and an increased protein content in bean plants treated with 0.5 $mg\,l^{-1}$ *atrazine*, *simazine*, or *terbutryne* has been reported [336, 372]. Clearly, then, the flow of freshly fixed carbon can be changed in the adapted plants with the result that a higher percentage of it is converted into nitrogenous products. Other increases, like e.g. for K^+, HPO_4^{2-}, Fe^{2+}, and Ca^{2+} [317, 337, 361], for RNA [319], and for several enzyme activities [346] may be termed tertiary effects. Their very distant relation to the primary inhibition may also be concluded from the fact that in other systems decreases of the same parameters

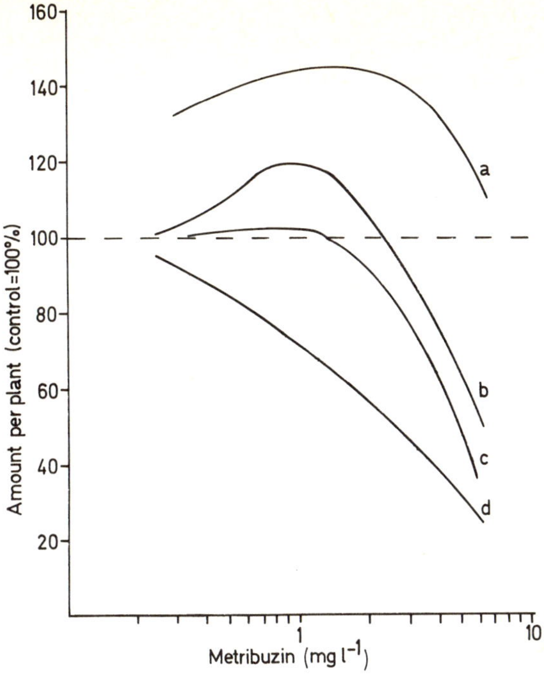

Fig. 31. Concentration dependence of compositional changes in potato plants grown at different concentrations of *metribuzin* [316]. *a* soluble amino acids, *b* soluble protein, *c* dry weight, *d* soluble reducing sugars

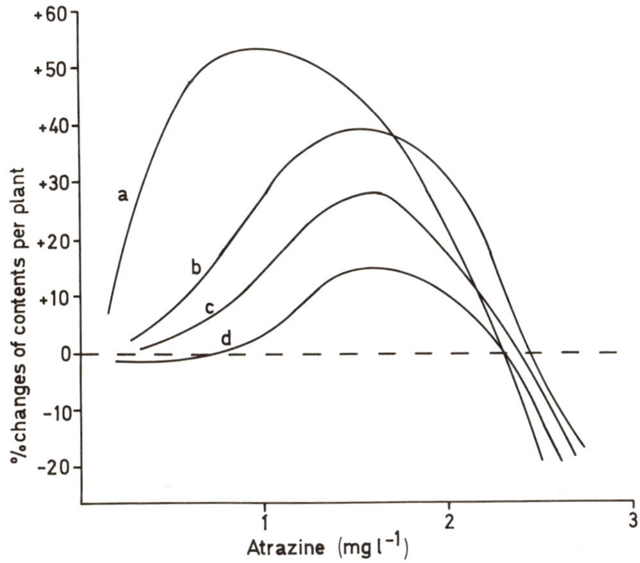

Fig. 32. Concentration dependence of compositional changes in maize plants grown at different concentrations of *atrazine* [317]. *a* chlorophyll- and lipid-N, *b* protein- and nucleic acid-N, *c* total N, *d* nitrate-N

may equally occur [137, 373, 374]. Truelove et al. [375] observed inhibitory effects caused by *prometryne*; however, they used elevated concentrations (10–100 μM). At these concentrations, *prometryne* not only inhibits photosynthesis but also uncouples oxidative phosphorylation [75, 114].

The delayed senescence mentioned before has been observed upon application of many different photosynthesis-inhibiting herbicides, e.g. of *atrazine*, *bromacil*, and *fluometuron* [186] and of *methabenzthiazuron* [310]. This effect may either be a consequence of changed endogenous hormone levels or simply of a developmental delay caused by the temporarily lowered carbohydrate supply, or both.

Whether the hormone activity of *s*-triazine herbicides (p. 76) contributes to a possible increase in yield is not clear. Several cases where the protein content was unchanged while the yield was increased have been reported. This could equally be the consequence of a hormone-type activity of the herbicide or of a "fertilizer effect" originating from increased nitrate uptake during the phase of a "shade adaptation". Unfortunately, the reproducibility of these effects under field conditions is very poor.

A further aspect is added to the complexity of regulatory and physiological responses by the observation that the sensitivity of plants treated with photosynthesis-inhibiting herbicides towards microbial infection is changed [376–379].

A close correlation between endogenous sugar levels and microbial growth rates is reflected by the fact that after *simazine*, *monolinuron*, or *methabenzthiazuron* treatment the infection caused by the "high sugar deseases" *Cercosporella herpotrichoides* and *Erysiphe graminis* is significantly decreased. Several weeks after application of the herbicide, however, an increase may occur. Conversely, the infection and growth rates of the so-called "low-sugar deseases", e.g. *Alternaria*, *Fusarium*, and *Helminthosporium (=Drechslera)*, might be decreased after partial inhibition of photosynthesis by these herbicides.

3. Hormone-Like Effects of *s*-Triazine Herbicides

As repeatedly referred to earlier, *s*-triazine herbicides display plant hormone-like activity in a number of assay systems (Table 21). The tests used most frequently, the callus growth test and the test measuring the delay of chlorophyll degradation, indicate a cytokinin-type activity. The *s*-triazine analogues with the highest cytokinin activity were *ametryne* and *prometryne; simazine* and *atrazine* displayed little or no activity in these systems (Table 22). However, the question of the specificity of these test systems must be raised. The callus growth test [387–390] and especially the chlorophyll retention test [391, 392] show rather varied specificities, also including e.g. phenylpyridyl-ureas. The structure-activity correlations of these ureas are, however, different from those of the photosynthesis-inhibiting phenylureas [392]. Other authors have been unable to reproduce a stimulation of callus growth although the total nitrogen content was increased [382]. Bush et al. [314] found an increased rate of protein synthesis in red kidney bean embryonic axis.

Another test indicative of cytokinin activity is the hypocotyl hook-opening test. Here again, *ametryne* and *prometryne* were the most active *s*-triazines whereas *atrazine* showed no effect. Soybean callus root-initiation experiments revealed

Table 21. Hormone-like effects of *s*-triazine herbicides (↑ increase or ↓ decrease)

Test system	Herbicide(s)	Author(s)	Concentration (M)	Effect
Sorghum callus growth	Ametryne Prometryne Propazine etc.	[380]	10^{-6}	↑
Soybean callus growth	Atrazine	[381]	$10^{-10}-10^{-18}$	↑
	Ametryne Prometryne	[382]	$10^{-5}-10^{-8}$	↓ ↑
Tobacco callus growth	Simazine Atrazine	[383]	10^{-6}	↓
Soybean callus root initiation (±IAA, Kinetin)	Atrazine Ametryne	[384]	10^{-14}	↑ ↓
Chromatin activity from etiolated soybean seedlings	Atrazine	[131]	$10^{-4}-10^{-6}$	↑
Cotton hypocotyl hook opening	Prometryne Ametryne etc.	[385]	10^{-8}	↑
Bush bean embryonic axis elongation	Atrazine	[314]	10^{-8}	↑
Chlorophyll degradation	Ametryne Prometryne	[380]	$2\times10^{-5}-2\times10^{-3}$	↓
	Atrazine	[386]	$5\times10^{-8}-10^{-7}$	↓ ↑
Lactuca germination	Ametryne Prometryne etc.	[385]	$10^{-8}-10^{-12}$	↑
IAA peroxidase	Atrazine	[381]	$10^{-6}-10^{-8}$ $10^{-10}-10^{-21}$	↑ ↓

Table 22. Promotion of sorghum callus growth by *s*-triazine herbicides and zeatin at 0.5 mg l^{-1} and by DNA nucleotides at 1 mg l^{-1} [380]

Addition	Callus weight (g/tube)
Control	0.28
Ametryne	0.81
Prometryne	0.63
Prometon	0.56
Propazine	0.47
Simazine	0.42
Atrazine	0.40
Zeatin	0.95
Guanine	0.38
Adenine	0.28
Thymidine	0.27
Cytosine	0.27

highly significant interactions between kinetin, IAA, and *s*-triazines. Other test systems showing possible *s*-triazine interactions with auxins involved the elongation of bean embryonic axis and the activity of IAA-peroxidase.

Tests operating with concentrations in excess of 10 μM may not be relevant to growth-stimulating effects of *s*-triazines at herbicidal rates in the field. However, the clearly demonstrated cytokinin activity of some *s*-triazines at micromolar concentrations points to the possible involvement of cytokinin activity of *s*-triazines in the observed growth stimulations. Recently, a new experimental approach with *s*-triazine-resistant weeds has become possible, and future work might answer some of the open questions.

References for C.I–VIII

1. Trebst, A.: Inhibitors of electron flow: tools for the functional and structural localization of carriers and energy conservation sites. In: Methods in Enzymology. San Pietro, A. (ed.), Vol. 69, p. 675. New York, London: Academic Press 1980
2. Trebst, A.: Action mechanism of herbicides in photosynthetic electron transport. V. Int. Photosynthesis Congr., Greece 1980
3. Krinsky, N.I.: Carotenoid pigments: multiple mechanisms for coping with the stress of photosensitized oxidations. In: Strategies of microbial life in extreme environments. Shilo, M. (ed.), p. 163. Weinheim, New York: Verlag Chemie 1979
4. Hassan, H.M., Fridovich, I.: Superoxide dismutase and its role for survival in the presence of oxygen. In: Strategies of microbial life in extreme environments. Shilo, M. (ed.), p. 179. Weinheim, New York: Verlag Chemie 1979
5. Halliwell, B.: Oxygen-free-radicals in living systems: dangerous but useful? In: Strategies of microbial life in extreme environments. Shilo, M. (ed.), p. 195. Weinheim, New York: Verlag Chemie 1979

5a. Elstner, E.F., Osswald, W., Konze, J.R.: FEBS Lett. *121*, 219 (1980)

5b. Nakano, Y., Asada, K.: Plant Cell Physiol. *21*, 1295 (1980)

6. Halliwell, B.: Prog. Biophys. Molec. Biol. *33*, 1 (1978)
7. Winterbourn, C.C.: Biochem. J. *182*, 625 (1979)
8. Heber, U.: Biochim. Biophys. Acta *305*, 140 (1973)
9. Marsho, T.V., Behrens, P.W., Radmer, R.J.: Plant Physiol. *64*, 656 (1979)
10. Elstner, E.F.: Oxygen activation and superoxide dismutase in chloroplasts. In: Photosynthesis II. Gibbs, M., Latzko, E. (eds.). Encyclopedia in Plant Physiol. New Ser. Vol. 6, p. 410. Berlin, Heidelberg, New York: Springer 1979
11. van Hasselt, P.P. et al.: Physiol. Plant. *45*, 475 (1979)
12. Beddard, G.S., Davidson, R.S., Trethewey, K.R.: Nature *267*, 373 (1977)
13. Searle, G.F.W., Wessels, J.S.C.: Biochim. Biophys. Acta *504*, 84 (1978)
14. Takahama, U.: Plant Cell Physiol. *19* (1978)
15. Farrington, J.A. et al.: Biochim. Biophys. Acta *314*, 372 (1973)
16. Fridovich, I.: Science *201*, 875 (1978)
17. Groden, D., Beck, E.: Biochim. Biophys. Acta *546*, 426 (1979)
18. Hager, A.: Ber. Deutsch. Bot. Ges. *88*, 27 (1975)

18a. Jeffrey, S.W., Douce, R., Benson, A.A.: Proc. Natl. Acad. Sci. U.S.A. *71*, 807 (1974)

18b. Siefermann, D., Yamamoto, H.Y.: Biochem. Biophys. Res. Comm. *62*, 456 (1975)

19. Good, N.E.: Plant. Physiol. *36*, 788 (1961)
20. Moreland, D.E., Hill, K.L.: J. Agr. Food Chem. *7*, 832 (1959)
21. Tischer, W., Strotmann, H.: Biochim. Biophys. Acta *460*, 113 (1977)
22. Renger, G.: Biochim. Biophys. Acta *314*, 113 (1973)
23. Izawa, S., Good, N.E.: Inhibition of photosynthetic electron transport and photophosphorylation. In: Methods in Enzymol. San Pietro, A. (ed.), Vol. XXIV, p. 355. New York, London: Academic Press 1972
24. Izawa, S.: Inhibitors of electron transport. In: Photosynthetic electron transport and photophosphorylation. Trebst, A., Avron, M. (eds.). Encyclopedia in Plant Physiol. New Ser. Vol. 5, p. 266. Berlin, Heidelberg, New York: Springer 1977
25. Moreland, D.E.: Ann. Rev. Plant Physiol. *31*, 597 (1980)
26. Bedell, G.W., Govindjee: Plant Cell Physiol. *14*, 1081 (1973)

27. Trebst, A., Reimer, S.: Reversal of the inhibition of photosynthetic electron flow in chloroplasts by an internal TMPD bypass. In: Photosynthetic organelles. Plant Cell Physiol. Special Issue, p. 201 (1977)
28. Trebst, A. et al.: Z. Naturforsch. *33 c*, 919 (1978)
29. Trebst, A., Draber, W.: Structure activity correlations of recent herbicides in photosynthetic reactions. In: Advances in pesticide science. Geissbühler, H. (ed.), p. 223. Oxford, New York: Pergamon Press 1979
30. Trebst, A., Harth, E.: Z. Naturforsch. *29 c*, 232 (1974)
31. Merbach, W., Schilling, G.: Biochem. Physiol. Pflanzen *171*, 187 (1977)
32. Hatzios, K.K., Penner, D.: Weed Res. *19*, 371 (1979)
33. Hatzios, K.K., Penner, D.: Weed Sci. *28*, 285 (1980)
34. Hatzios, K.K., Penner, D.: Pestic. Biochem. Physiol. *13*, 237 (1980)
35. Hatzios, K.K., Penner, D., Bell, D.: Plant Physiol. *65*, 319 (1980)
36. Hatzios, K.K., Penner, D.: Weed Sci. *28*, 97 (1980)
37. Dicks, I.W.: Pestic. Sci. *9*, 59 (1978)
38. Boulware, M.A., Camper, N.D.: Physiol. Plant. *26*, 313 (1972)
39. Boulware, M.A., Camper, N.D.: Weed Sci. *21*, 145 (1973)
40. Moreland, D.E.: Inhibitors of chloroplast electron transport: structure-activity relations. In: Progress in photosynthesis research. Metzner, H. (ed.), p. 1693. Tübingen: Laupp Press 1969
41. Wilkinson, R.E., Hardcastle, W.S.: J. Sci. Food Agr. *23*, 1301 (1972)
42. Philippines Int. Rice Res. Inst.: High protein rice. In: 1971 annual report, p. 165 (1972)
43. Müller, F., Frahm, J., Sanad, A.: Weed Abstr. *27*, 333 (1978)
44. Holm, R.E., Stallard, D.E.: Weed Sci. *22*, 10 (1974)
45. Glasgow, J.L., Dicks, J.W.: Weed Res. *20*, 17 (1980)
46. Hammerschlag, R.S. et al.: Weed Sci. *23*, 425 (1975)
47. Nishimura, K. et al.: Agr. Biol. Chem. *39*, 867 (1975)
48. Suzuki, R., Fujita, Y.: Plant Cell Physiol. *18*, 625 (1977)
49. Codd, G.A., Cossar, J.D.: Biochem. Biophys. Res. Comm. *83*, 342 (1978)
50. Moreland, D.E., Boots, M.R.: Plant Physiol. *47*, 53 (1971)
51. van Rijn, P.J.: Weed Res. *12*, 284 (1972)
52. Baldwin, R.C.: Weed Abstr. *22*, 19 (1973)
53. Prius, H.B.A.: Acta Bot. Neerl. *19*, 813 (1970)
54. Posner, H.B., Posner, R.S., Gower, R.A.: Plant Cell Physiol. *18*, 1301 (1977)
55. Löppert, H., Kronberger, W., Kandeler, R.: Planta *138*, 133 (1978)
56. Imhoff, C.: Comt. Rend. Ser. D *276*, 3303 (1973)
57. Lay, M.M., Ilnicki, R.D.: Weed Abstr. *22*, no. 2970 (1973)
58. Lay, M.M., Ilnicki, R.D.: Weed Abstr. *22*, no. 1833 (1973)
59. van Oorschot, J.L.P.: Acta Bot. Neerl. *23*, 36 (1974)
60. van Oorschot, J.L.P.: Pestic. Sci. *1*, 33 (1970)
61. Sumida, S., Yoshida, R., Ueda, M.: Plant Cell Physiol. *16*, 257 (1975)
62. Göbel, E., Riessner, R., Pohl, P.: Z. Naturforsch. *31 c*, 687 (1976)
63. Orr, A.R., Kessler, J.E., Tepaske, E.R.: Amer. J. Bot. *63*, 973 (1976)
64. Watanabe, M., Miyoshi, Y., Furuya, M.: Plant Cell Physiol. *17*, 683 (1976)
65. Jeanjean, R.: Physiol. Plant. *37*, 107 (1976)
66. Inoue, Y., Ishizuka, K., Mitsu, S.: Agr. Biol. Chem. *34*, 1029 (1970)
67. Deysson, G., Thizy, A., Neguleseo-Rubingher, I.: Weed Abstr. *25*, no. 320 (1976)
68. Liu, C.Y. et al.: Plant Physiol. Suppl. *49*, no. 60 (1972)
69. Convent, B., Briquet, M.: Weed Abstr. *27*, no. 3097 (1978)
70. Boots, M.V., Boots, S.G., Moreland, D.E.: J. of Med. Chem. *13*, 144 (1970)
71. Walsh, G.E., Grow, T.E.: Weed Sci. *19*, 568 (1971)
72. Sikka, H.C., Pramer, D.: Weed Sci. *16*, 296 (1968)
73. Wergin, W.P., Potter, J.R.: Pestic. Biochem. Physiol. *5*, 265 (1975)
74. Feierabend, J., Schubert, B.: Plant Physiol. *61*, 1017 (1978)
75. McDaniel, J.L., Frans, R.E.: Weed Sci. *17*, 129 (1969)
76. Rubin, B., Eshel, J.: Phytoparasitica *4*, 142 (1976)
77. Sweeney, J.P., Marsh, A.C.: J. Agr. Food Chem. *19*, 854 (1971)
78. Salminen et al.: Acta Agaric. Scand. *20*, 49 (1970)

79. Amberger, A., Gutser, R.: Z. Pflanzenern. Bodenk. *138*, 337 (1975)
80. Perry, M.W.: Weed Res. *13*, 325 (1973)
81. Walker, A., Featherstone, R.E.: Weed Abstr. *22*, no. 1842 (1973)
82. Muschinek, G. et al.: Weed Res. *19*, 101 (1979)
83. Butts, E.G., Foy, C.L.: Pestic. Biochem. Physiol. *4*, 44 (1974)
84. Dorough, H.W., Whitacre, D.M., Cardona, R.A.: J. Agr. Food Chem. *21*, 797 (1973)
85. Dorough, H.W.: Bull. Environ. Contam. and Toxicol. *12*, 493 (1974)
86. Keeley, P.E., Thullen, R.J.: Pestic. Biochem. Physiol. *10*, 275 (1979)
87. Luzhnova, M.I., Shekhtman, L.M.: Pestic. Biochem. Physiol. *5*, 205 (1975)
88. Pusztai, T., Vegh, A.: Weed Abstr. *29*, no. 838 (1980)
89. Lee, S.S., Fang, S.C.: Phytochem. *11*, 2693 (1972)
90. Minshall, W.H.: Can. J. Bot. *38*, 201 (1960)
91. Malakondaiah, N., Fang, S.C.: Pestic. Biochem. Physiol. *9*, 33 (1978)
92. Moreland, D.E., Hill, K.L.: Weeds *10*, 229 (1962)
93. Hollister, T., Walsh, G.E.: Bull. Environ. Contam. and Toxicol. *9*, 291 (1973)
94. Ducruet, J.M., Gauvrit, C.: Weed Res. *18*, 327 (1978)
95. Gauvrit, C.: Physiol. Plant. *48*, 448 (1980)
96. Splitstoesser, W.E., Hopen, H.J.: Physiol. Plant. *23*, 964 (1970)
97. Lee, I.N., Ishizuka, K.: Arch. Environ. Contam. and Toxicol. *4*, 155 (1976)
98. Alsop, W.R., Moreland, D.E.: Pestic. Biochem. Physiol. *5*, 163 (1975)
99. Shirakawa, N., Ueki, K.: Weed Abstr. *26*, 92 (1977)
100. Mine, A., Matsunaka, S.: Pestic. Biochem. Physiol. *5*, 444 (1975)
101. Wakamori, S.: Agr. Biol. Chem. *37*, 307 (1973)
102. Bethlenfalvay, G.J., Castelfranko, P.A.: Weed Sci. *26*, 84 (1978)
103. Hofstra, G., Switzer, C.M.: Weed Sci. *16*, 23 (1968)
104. Gruenhagen, R.D., Moreland, D.E.: Weeds *19*, 319 (1971)
105. Yih, R.Y., Mc Rae, D.H., Wilson, H.F.: Plant Physiol. *43*, 1291 (1968)
106. Hoagland, R.E., Graf, G.: Weed Sci. *20*, 303 (1972)
107. Hoagland, R.E., Graf, G., Handel, E.D.: Weed Res. *14*, 371 (1974)
108. Hoagland, R.E., Graf, G.: Phytochem. *11*, 521 (1972)
109. Hoagland, R.E.: Phytochem. *14*, 383 (1975)
110. Hoagland, R.E.: Plant Cell Physiol. *19*, 1019 (1978)
111. Yukimoto, M., Oda, M.: Weed Abstr. *23*, no. 1629 (1974)
112. Sumida, S., Ueda, M.: Plant Cell Physiol. *14*, 781 (1973)
113. Rau, I., Grimme, L.H.: Z. Naturforsch. *26b*, 919 (1971)
114. Thompson, O.C., Truelove, B., Davis, D.E.: Weed Sci. *22*, 164 (1974)
115. Thompson, O.C., Truelove, B., Davis, D.E.: J. Agr. Food Chem. *17*, 997 (1969)
116. Gysin, H., Knüsli, E.: Chemistry and herbicidal properties of triazine derivatives. In: Advances in pest control research. Metcalf, R.L. (ed.), p. 289. New York: Interscience Publ. 1960
117. Imbamba, S.K., Moss, D.N.: Crop Sci. *11*, 844 (1971)
118. Ladlie, J.S., Meggit, W.F., Penner, D.: Weed Sci. *25*, 115 (1977)
119. Bolhar-Nordenkampf, H.R.: Biochem. Physiol. Pfl. *169*, 121 (1976)
120. Hill, E.R., Patula, E.C., Vulgris, J.: Weed Sci. *16*, 377 (1968)
121. Lamoureux, G.L. et al.: J. Agr. Food Chem. *21*, 1020 (1973)
122. Shimabukuro, R.H., Swanson, H.R.: J. Agr. Food Chem. *17*, 199 (1969)
123. Shimabukuro, G.L. et al.: Plant Physiol. *46*, 103 (1970)
124. Shimabukuro, G.L. et al.: Plant Physiol. *47*, 10 (1971)
125. Shimabukuro, G.L. et al.: J. Agr. Food Chem. *21*, 1031 (1973)
126. Guddewar, M.B., Dautermann, W.C.: Phytochem. *18*, 735 (1979)
127. Liang, G.H., Liang, Y.T.S.: Can. J. Genet. Cytol. *14*, 423 (1972)
128. Müller, A., Ebert, E., Gast, A.: Experientia *28*, 704 (1972)
129. Mücke, W., Müller, P.W., Esser, H.O.: Experientia *25*, 353 (1969)
130. Plewa, M.J., Gentile, J.M.: Weed Abstr. *27*, no. 2707 (1978)
131. Penner, D., Early, R.W.: Weed Sci. *20*, 365 (1972)
132. Foy, C.L., Penner, D.: Weeds *13*, 226 (1965)
133. Jensen, K.J.N. et al.: Weed Res. *17*, 379 (1977)
134. Büchel, K.H.: Pestic. Sci. *3*, 89 (1970)

135. Ashton, F.M., Glenn, R.K.: Pestic. Biochem. Physiol. *11*, 201 (1979)
136. Sikka, H.C., Davis, D.E.: Weed Sci. *17*, 122 (1969)
137. Whitenberg, D.C., Flippen, J.H.: Weed Abstr. *27*, no. 3431 (1978)
138. Davis, D.E., Pillai, C.G.P., Truelove, B.: Weed Sci. *24*, 587 (1976)
139. Berüter, J., Temperli, A.T.: Experientia *26*, 600 (1970)
140. Temperli, A., Türler, H., Ercegovichi, C.D.: Z. Naturforsch. *21b*, 903 (1966)
141. Singh, J.N., Basler, E., Santelmann, P.W.: Pestic. Biochem. Physiol. *2*, 143 (1972)
142. Gast, A.: Experientia *14*, 134 (1957)
143. Shaukat, S.S., Moore, K.G., Lovell, P.H.: Physiol. Plant. *33*, 295 (1975)
144. Sutton, D.L. et al.: Weed Sci. *17*, 56 (1969)
145. Thompson, L.: Weed Sci. *20*, 153 (1972)
146. Thompson, L.: Weed Sci. *20*, 584 (1972)
147. Gnanarethinam, J.L.: Weed Res. *15*, 143 (1975)
148. Tieszen, L.L.: Plant Physiol. *46*, 442 (1970)
149. Smith, A.E., Wilkinson, R.E.: Weed Sci. *21*, 57 (1973)
150. van Oorschot, J.L.P.: Weed Res. *14*, 75 (1974)
151. van Oorschot, J.L.P.: Weed Res. *14*, 81 (1974)
152. Rubin, B.: Weed Abstr. *28*, no. 1676 (1979)
153. Rubin, B., Eshel, Y.: Weed Abstr. *27*, no. 515 (1978)
154. Eshel, Y., Kovacs, M., Rubin, B.: Pestic. Biochem. Physiol. *5*, 295 (1975)
155. Figuerola, L.F., Furtick, W.R.: Weed Sci. *20*, 60 (1972)
156. Houseworth, L.D., Tweedy, B.G.: Weed Sci. *19*, 732 (1971)
157. Schmidt, R.R. et al.: Pestic. Sci. *6*, 239 (1975)
158. Schmidt, R.R., Fedtke, C.: Pestic. Sci. *8*, 611 (1977)
159. Müller, F., Sanad, A.: Med. Fac. Landbouww. Rijksuniv. Gent *41*, 1211 (1976)
160. van Oorschot, J.L.P., van Leeuwen, P.H.: Weed Res. *19*, 63 (1979)
161. Schmidt, R.R.: Med. Fac. Landbouww. Rijksuniv. Gent *41*, 1229 (1976)
162. Fedtke, C., Schmidt, R.R.: Z. Naturforsch. *34c*, 948 (1979)
163. Fedtke, C.: unpublished
164. Draber, W., Dickoré, K., Büchel, K.H.: Naturwiss. *55*, 446 (1968)
165. Draber, W. et al.: Structure activity correlation of 1,2,4-triazinones, a new group of photosynthesis inhibitors. In: Progress in photosynthesis research. Metzner, H. (ed.), Vol. III, p. 1789. Tübingen: Laupp Press 1969
166. Trebst, A., Wietoska, H.: Z. Naturforsch. *30c*, 499 (1975)
167. Schmidt, R.R.: Z. PflKrankh. PflSchutz, Sonderh. VII, 109 (1975)
168. Fortino, J., Splitstoesser, J., Splitstoesser, W.E.: Weed Sci. *22*, 460 (1974)
169. Phatak, S.C., Stephenson, G.R.: Can. J. Plant Sci. *53*, 843 (1973)
170. Pritchard, M.K., Warren, G.F.: Weed Sci. *28*, 186 (1980)
171. Graf, G.T., Ogg, A.G.: Weed Sci. *24*, 137 (1976)
172. Hardcastle, W.S., Wilkinson, R.E., Young, C.T.: Weed Sci. *22*, 575 (1974)
173. Hargroder, T.G., Rogers, R.L.: Weed Sci. *22*, 238 (1974)
174. Hardcastle, W.S.: Weed Res. *14*, 181 (1974)
175. Hardcastle, W.S.: Weed Sci. *27*, 278 (1979)
176. da Silva, J.F.: Weed Abstr. *26*, no. 1779 (1977)
177. Smith, A.E., Wilkinson, R.E.: Physiol. Plant. *32*, 253 (1974)
178. Stephenson, G.R., McLeod, J.E., Phatak, S.C.: Weed Sci. *24*, 161 (1976)
179. Gaeronnski, S.W., Callihan, R.H., Pavek, J.J.: Weed Sci. *25*, 122 (1977)
180. Coble, H.D., Schrader, J.W.: Weed Sci. *21*, 308 (1973)
181. Oswald, T.H., Smith, A.E., Phillips, D.V.: Pestic. Biochem. Physiol. *8*, 73 (1978)
182. Mann, M.A., Mc Leod, W.J.: Can. J. Plant Sci. *58*, 485 (1978)
183. Edwards, C.J., Barrentine, W.L., Kilen, T.C.: Crop Sci. *16*, 119 (1976)
184. Rosen, J.D., Siewierski, M.: Bull. Environ. Contam. and Toxicol. *6*, 406 (1971)
185. Hoffman, C.E.: The mode of action of bromacil and related uracils. In: Proc. II. Int. IUPAC Congr. of Pestic. Chem., Vol. V. Tahori, A.S. (ed.), p. 65. New York, London, Paris: Gordon and Breach 1972
186. Hiranpradit, H., Foy, C.L.: Bot. Gaz. *134*, 26 (1973)
187. Shriver, J.W., Bingham, S.W.: Weed Sci. *21*, 212 (1973)

188. Weierich, A.J., Nelson, Z.A., Appleby, A.P.: Weed Sci. *25*, 27 (1977)
189. Bethlenfalvay, G., Norris, R.F.: Weed Sci. *23*, 499 (1975)
190. Bethlenfalvay, G., Norris, R.F.: Weed Sci. *25*, 407 (1977)
191. Trebst, A. et al.: Z. Naturforsch. *23b*, 342 (1968)
192. Arndt, F., Kötter, C.: Weed Res. *8*, 259 (1968)
193. Kötter, C., Arndt, F.: Z. PflKrankh. PflSchutz, Sonderh. V, 81 (1970)
194. Kassebeer, H.: Z. PflKrankh. PflSchutz *78*, 158 (1971)
195. St. John, J.B., Hilton, J.L.: Weed Sci. *24*, 579 (1976)
196. Frank, R., Switzer, C.M.: Weed Sci. *17*, 344 (1969)
197. Irbe, K.J., Romanovskaya, O.J.: Weed Abstr. *21*, no. 2540 (1972)
198. Strang, R.H., Rogers, R.L.: Weed Sci. *23*, 26 (1975)
199. Strang, R.H., Rogers, R.L.: J. Agr. Food Chem. *22*, 1119 (1974)
200. Eshel, Y.: Weed Res. *9*, 167 (1969)
201. Würzer, B.: Naturwiss. *56*, 452 (1969)
202. Fischer, A.: Weed Res. *2*, 177 (1962)
203. Frank, R., Switzer, C.M.: Weed Sci. *17*, 365 (1969)
204. Reimer, S., Link, K., Trebst, A.: Z. Naturforsch. *34c*, 419 (1979)
205. Moreland, D.E., Blackmon, W.J.: Weed Sci. *18*, 419 (1970)
206. Katoh, S.: Plant Cell Physiol. *13*, 273 (1972)
207. Savory, B.M., Hibbitt, C.J., Catchpole, A.H.: Pestic. Sci. *6*, 143 (1975)
208. Schafer, D.E., Chilcote, D.O.: Weed Sci. *18*, 729 (1970)
209. Rost, T.L., Morrison, S.L., Sachs, E.S.: Amer. J. Bot. *64*, 780 (1977)
210. Robert, M.L., Taylor, H.F., Wain, R.L.: Planta *129*, 53 (1976)
211. Larque-Saavedra, A., Wilkings, H., Wain, R.L.: Planta *126*, 269 (1975)
211a. Wain, R.L., Whitford, P.N.: Ann. Appl. Biol. *96*, 255 (1980)
212. van Rensen, J.J.S., Wong, D., Govindjee: Z. Naturforsch. *33c*, 413 (1978)
213. van Rensen, J.J.S., van der Vet, W., van Vliet, W.P.A.: Photochem. Photobiol. *25*, 579 (1977)
214. Röchling, H., Büchel, K.H.: Chem. Ber. *104*, 344 (1971)
215. Büchel, K.H. et al.: Angew. Chemie *77*, 911 (1965)
216. Trebst, A., Pistorius, E., Büchel, K.H.: Z. Naturforsch. *21b*, 667 (1966)
217. Bond, C.P., Corbett, J.R.: Biochem. J. *118*, 501 (1970)
218. Bond, C.P., Corbett, J.R.: Pestic. Sci. *2*, 169 (1971)
219. Retzlaff, G., Fischer, A.: Weed Abstr. *23*, no. 1633 (1974)
220. Pfister, K., Buschmann, C., Lichtenthaler, H.K.: Inhibition of photosynthetic electron transport by bentazon. In: Proc. III. Int. Photosynthesis Congr. Avron, M. (ed.), p. 675 (1974)
221. Potter, J.R., Wergin, W.P.: Pestic. Biochem. Physiol. *5*, 458 (1975)
222. Böger, P., Beese, B., Miller, R.: Weed Res. *17*, 61 (1977)
223. Retzlaff, G., Hilton, J.L., St. John, J.B.: Z. Naturforsch. *34c*, 944 (1979)
224. Potter, J.R.: Weed Sci. *25*, 241 (1977)
225. Weber, J.B.: Weed Sci. *28*, 467 (1980)
226. Retzlaff, G., Hamm, R.: Proc. 41. Deutsche Pflanzenschutztagg., p. 220 (1977)
227. Mine, A., Miyakado, M., Matsunaka, S.: Pestic. Biochem. Physiol. *5*, 566 (1975)
228. Dannigkeit, W.: Z. PflKrankh. PflSchutz *84*, 540 (1977)
229. Retzlaff, G., Hamm, R.: Weed Res. *16*, 263 (1976)
230. Duncan, D.N., Meggit, W.F., Penner, D.: Proc. Weed Sci. Soc. Amer. Conf., no. 171 (1977)
231. Trebst, A., Harth, E., Draber, W.: Z. Naturforsch. *25b*, 1157 (1970)
232. Oettmeier, W., Reimer, S., Link, K.: Z. Naturforsch. *33c*, 695 (1978)
233. Siedow, J.N., Huber, S.C., Moreland, D.E.: Biochim. Biophys. Acta *547*, 282 (1979)
234. Zweig, G., Hitt, J.E., Cho, D.H.: J. Agr. Food Chem. *17*, 176 (1969)
235. Sikka,, H.C., Carroll, J., Zweig, G.: Pestic. Biochem. Physiol. *1*, 381 (1971)
236. Sikka, H.C., Shimabukuro, R.H., Zweig, G.: Plant. Physiol. *49*, 381 (1972)
237. Sikka, H.C., Saxena, J., Zweig, G.: Plant. Physiol. *51*, 363 (1973)
238. Schwelitz, F.D. et al.: Pestic. Biochem. Physiol. *4*, 379 (1974)
239. Oettmeier, W., Masson, K.: Pestic. Biochem. Physiol. *14*, 86 (1980)
240. Oettmeier, W., Masson, K.: FEBS Lett. *118*, 267 (1980)
240a. Oettmeier, W., Masson, K., Johanningmeier, U.: Biochim. Biophys. Acta, in press
240b. Steinback, K.E., Pfister, K., Arntzen; C.J.: Z. Naturforsch. *36c*, 98 (1981)

240c. Pfister, K.: in preparation
241. Pfister, K., Steinback, K.E., Arntzen, C.J.: Characterization of the thylacoid membrane protein which binds photosystem II inhibitors, V. Int. Photosynthesis Congr., Greece 1980
242. Pfister, K. et al.: Proc. Natl. Acad. Sci.U.S.A. *78*, 981 (1981)
243. Shipman, L.L.: J. of theor. Biol., in press
244. van Rensen, J.J.S., Vermaas, W.F.J.: Interaction of the bicarbonate ion with photosystem II inhibiting herbicides in photosynthetic electron transport, V. Int. Photosynthesis Congr., Greece 1980
245. Vermaas, W.F.J., van Rensen, J.J.S.: The mechanism of bicarbonate action in isolated broken chloroplasts, V. Int. Photosynthesis Congr., Greece 1980
245a. Khanna, R. et al.: Biochim. Biophys. Acta *634*, 105 (1981)
246. Renger, G.: Biochim. Biophys. Acta *440*, 287 (1976)
247. Trebst, A.: Z. Naturforsch. *34c*, 986 (1979)
248. Renger, G.: Z. Naturforsch. *34c*, 1010 (1979)
249. Böger, P., Kunert, K.-J.: Z. Naturforsch. *34c*, 1015 (1979)
250. Edelman, M., Mattoo, A.K., Hoffmann-Falke, H.: Regulation and identification of translation products in the chloroplast, V. Int. Photosynthesis Congr., Greece 1980
251. Bogorad, L.: Maps models and expression of the chloroplast genome. V. Int. Photosynthesis Congr., Greece 1980
252. Jordan, B.R., Harwood, J.L.: Biochem. J. *191*, 791 (1980)
253. Lehoczki, E., Herczog, T., Szalay, L.: Biochim. Biophys. Acta *545*, 376 (1979)
254. Herczog, T. et al.: Z. Pflanzenphysiol. *94*, 55 (1979)
255. LeBaron, H., Gressel, J. (eds.): Herbicide resistance in plants. J. Wiley Inc., in press
255a. Aigle, N.: La Defense des Vegetaux *34*, 323 (1980)
256. van Oorschot, J.L.P.: Z. Naturforsch. *34c*, 900 (1979)
257. Gressel, J.: Genetic herbicide resistance; projections on appearence in weeds and breeding for it in crops. In: Plant regulation and world agriculture. Scott, T.K. (ed.), Plenum Press 1979
258. Gressel, J., Segel, L.A.: J. theor. Biol. *75*, 1 (1978)
259. Hensley, J.R., Counselman, C.J.: Proc. Weed Sci. Soc. Amer. Conf., no. 232 (1979)
259a. Holt, J.S., Stemler, A.J., Radosevich, S.R.: Plant Physiol. *67*, 744 (1981)
260. Radosevich, S.R., Appleby, A.P.: Weed Sci. *21*, 497 (1973)
261. Radosevich, S.R.: Weed Sci. *25*, 316 (1977)
262. Bandeen, J.D., McLaren, R.D.: Can. J. Plant Sci. *56*, 411 (1976)
263. Jensen, K.J.N.: Weed Res. *17*, 379 (1977)
264. Gasquez, J., Barralis, G.: Chemosphere *11*, 911 (1978)
265. Radosevich, S.R., Steinback, K.E., Arntzen, C.J.: Weed Sci. *27*, 216 (1979)
266. Radosevich, S.R., Devilliers, O.T.: Weed Sci. *24*, 229 (1976)
267. Arntzen, C.J., Ditto, C.L., Brewer, P.E.: Proc. Natl. Acad. Sci. U.S.A. *76*, 278 (1979)
268. Machado, V.S. et al.: Weed Sci. *26*, 318 (1978)
269. Ducruet, J.M., Gasquez, J.: Chemosphere *8*, 691 (1978)
270. Grignac, P.: Weed Abstr. *28*, no. 1065 (1979)
271. Miele, S., Vannozzi, G.P.: Weed Abstr. *27*, no. 3414 (1978)
272. Lien, S. et al.: Plant Cell Physiol., Special Issue No. 3: Photosynthetic Organelles 243 (1977)
273. Böger, P., Sandmann, G., Miller, R.: Photosynthesis Res. *2*, 61 (1981)
274. Laval-Martin, D., Dubertret, G., Calvayrac, R.: Plant Sci. Lett. *10*, 185 (1977)
275. Pfister, K., Arntzen, C.J.: Z. Naturforsch. *34c*, 996 (1979)
276. Pfister, K., Radosevich, S.R., Arntzen, C.J.: Plant Physiol. *64*, 995 (1979)
277. Draber, W., Fedtke, C.: Herbicide interaction with plant biochemical systems. In: Advances in pestic. sci., Part 3. Geissbühler, H. (ed.), p. 475. Oxford, New York: Pergamon Press 1979
278. Pillai, P., St. John, J.B.: Plant. Physiol. *68*, 585 (1981)
279. Cooke, A.R.: Weeds *4*, 397 (1956)
280. Wessels, J.S.C., van der Veen, R.: Biochim. Biophys. Acta *19*, 548 (1956)
281. Sweetser, P.B., Todd, C.W.: Biochim. Biophys. Acta *51*, 504 (1961)
282. Fedtke, C.: Pestic. Sci. *4*, 653 (1973)
283. Bolhar-Nordenkampf, H.: Biochem. Physiol. Pfl. *161*, 342 (1970)
284. Ashton, F.M., Gifford, E.M., Bisalputra, T.: Bot. Gaz. *124*, 336 (1963)
285. Hill, E.R., Putala, E.C., Vengris, J.: Weed Sci. *16*, 377 (1968)

286. Anderson, J.L., Thomson, W.W.: Resid. Rev. *47*, 167 (1973)
287. Moreland, D.E.: Mode of action of herbicides. In: ACS Symposium Series No. 37. Plimmer, J.R. (ed.), p. 56. American Chemical Soc. 1977
288. Klepper, L.A.: Weed Sci. *23*, 188 (1975)
289. Klepper, L.A.: Plant Physiol. *64*, 273 (1979)
290. Churchill, K., Klepper, L.A.: Pestic. Biochem. Physiol. *12*, 156 (1979)
291. Fedtke, C.: Pestic. Sci. *8*, 152 (1977)
292. Lawrence, J.M., Foster, R.J., Herrick, H.E.: Plant Physiol. *65*, 984 (1980)
293. Finke, R.L., Warner, R.L., Muzik, T.J.: Weed Sci. *25*, 18 (1977)
294. Jones, R.W., Sheard, R.W.: Plant Sci. Lett. *11*, 285 (1978)
295. Stanger, C.E., Appleby, A.P.: Weed Sci. *20*, 357 (1972)
296. Pallett, K.E., Dodge, A.D.: J. Exp. Bot. *31*, 1051 (1980)
297. Ridley, S.M.: Plant Physiol. *59*, 724 (1977)
298. Giannopolitis, C.N., Ayers, G.S.: Weed Sci. *26*, 440 (1978)
299. Elstner, E.F., Pils, I.: Z. Naturforsch. *34c*, 1040 (1979)
300. Pallett, K.E., Dodge, A.D.: Z. Naturforsch. *34c*, 1058 (1979)
301. Pallett, K.E., Dodge, A.D.: Proc. Brit. Crop. Prot. Conf. – Weeds, 235 (1976)
302. van Rensen, J.J.S., Vermaas, W.F.J.: Physiol. Plant. *51*, 106 (1981)
303. Brown, B.T.: A comparison of hill inhibition and herbicidal activity of N-3-alkoxyuracil derivatives. In: 4. Int. Congr. of Pestic. Chem. (IUPAC), Abstr. Vol., no. II–451. Zürich 1978
304. Ries, S.K.: Subtoxic effects on plants. In: Herbicides – Physiology, Biochemistry, Ecology, Vol. 2. Audus, L.J. (ed.), p. 313. London, New York, San Francisco: Academic Press 1976[2]
305. Eastin, E.F., Davis, D.E.: Weed Sci. *15*, 306 (1967)
306. Pulver, E.L., Ries, S.K.: Weed Sci. *21*, 233 (1973)
307. Tweedy, J.A., Ries, S.K.: Plant Physiol. *42*, 280 (1967)
308. Fedtke, C.: Weed Sci. *27*, 192 (1979)
309. Lawrence, J.M., Foster, R.J., Herrick, H.E.: Plant Physiol. *65*, 984 (1980)
310. Fedtke, C.: Pestic. Sci. *4*, 653 (1973)
311. Fedtke, C.: Pestic. Biochem. Physiol. *4*, 386 (1974)
312. Ries, S.K. et al.: Proc. Natl. Acad. Sci. U.S.A. *58*, 526 (1967)
313. Lichtenthaler, H.K. et al.: Photosynth. Res. *1*, 29 (1980)
314. Bush, P.B., Ries, S.K.: Weed Sci. *22*, 227 (1974)
315. Fedtke, C.: Ber. Deutsch. Bot. Ges. *87*, 155 (1974)
316. Fedtke, C.: Pestic. Biochem. Physiol. *2*, 312 (1972)
317. Hiranpradit, H., Foy, C.L.: Agron. J. *64*, 267 (1972)
318. Fedtke, C.: Z. Naturforsch. *34c*, 932 (1979)
319. Ebert, E., van Assche, C.J.: Weed Abstr. *20*, no. 365 (1971)
320. Beaumont, G., Bastin, R., Therrien, H.P.: Weed Abstr. *27*, no. 1087 (1978)
321. Meier, D., Lichtenthaler, H.K., Burkard, G.: Z. Naturforsch. *35c*, 656 (1980)
322. Lichtenthaler, H.K.: Regulation of prenylquinone synthesis in higher plants. In: Lipids and lipid polymers in higher plants. Tevini, M., Lichtenthaler, H.K. (eds.), p. 231. Berlin, Heidelberg, New York: Springer 1977
323. Schiff, J.A., Zeldin, M.H., Rubman, J.: Plant Physiol. *42*, 1716 (1967)
324. Böger, P., Schlue, U.: Weed Res. *16*, 149 (1976)
325. Fedtke, C., Deichgräber, G., Schnepf, E.: Biochem. Physiol. Pfl. *171*, 307 (1977)
326. Kleudgen, H.K.: Pestic. Biochem. Physiol. *9*, 57 (1978)
327. Fedtke, C.: Weed Sci. *27*, 192 (1979)
328. Klein, S., Neumann, J.: Plant Cell Physiol. *7*, 115 (1966)
329. Kleudgen, H.K.: Z. Naturforsch. *34c*, 110 (1979)
330. Janauer, G.A., Kinzel, H.: Z. Pflanzenphysiol. *78*, 52 (1976)
331. Janauer, G.A., Kinzel, H.: Z. Pflanzenphysiol. *77*, 383 (1976)
332. Janauer, G.A., Kinzel, H.: Biochem. Physiol. Pfl. *169*, 45 (1976)
333. Ries, S.K., Wert, V.: Weed Sci. *20*, 569 (1972)
334. Mohandas, S., Wallace, W., Nicholas, D.J.D.: Phytochem. *17*, 1021 (1978)
335. Singh, B., Salunkhe, D.K., Lipton, S.H.: The Journal of Hort Sci. *47*, 441 (1972)
336. Singh, B., Campbell, W.F., Salunkhe, D.K.: Amer. J. Bot. *59*, 568 (1972)
337. Singh, B. et al.: J. Agr. Food Chem. *20*, 1256 (1972)

338. Aslam, M., Huffaker, R.C.: Physiol. Plant. *28*, 400 (1973)
339. Aamisepp, A.: Weed Abstr. *25*, no. 3110 (1976)
340. Gramlich, J.V., Davis, D.E.: Weed Sci. *15*, 157 (1967)
341. Houston, W.R., van der Sluijs, D.H.: J. of Range Management *28*, 372 (1975)
342. Kay, B.L.: Weed Sci. *19*, 370 (1971)
343. Decleire, M., van Roey, G., Bastin, R.: Biochem. Physiol. Pfl. *170*, 443 (1976)
344. McCarty, M.K.: Weed Sci. *27*, 415 (1979)
345. Moyer, J.L., Paulsen, G.M.: Agron. J. *69*, 58 (1977)
346. Wu, M.T., Singh, B., Salunkhe, D.K.: J. Exp. Bot. *23*, 793 (1972)
347. Rottmann, G.A., Tweedy, J.A., Kapusta, G.: Weed Abstr. *24*, no. 2365 (1975)
348. Phatak, S.C., Cantliffe, D.J.: Hort. Sci. *10*, 271 (1975)
349. Cantliffe, D.J., Phatak, S.C.: Hort. Sci. *9*, 470 (1974)
350. Boardman, N.K.: Ann. Rev. Plant Physiol. *28*, 355 (1977)
351. Wild, A.: Ber. Deutsch. Bot. Ges. *92*, 341 (1979)
352. Tanaka, K., Kato, S., Yomo, S.: Plant Sci. Lett. *10*, 31 (1977)
353. Fedtke, C.: Nitrogen metabolism in photosynthetically inhibited plants. In: Biology of inorganic nitrogen and sulfur. Bothe, H., Trebst, A. (eds.), p. 260. Berlin, Heidelberg: Springer 1981
354. Purvis, A.C., Peters, D.B., Hageman, R.H.: Plant Physiol. *53*, 934 (1974)
355. Neyra, C.A., Hageman, R.H.: Plant Physiol. *58*, 726 (1976)
356. Aslam, M., Huffaker, R.C.: Physiol. Plant. *28*, 400 (1973)
357. Klepper, L., Flesher, D., Hageman, R.H.: Plant Physiol. *48*, 580 (1971)
358. Chodova, D., Zemanek, J.: Weed Abstr. *27*, no. 2705 (1978)
359. Dedatta, S.K., Obcemea, W.N., Jana, R.K.: Agron. J. *64*, 785 (1972)
360. Currey, W.L.: Weed Abstr. *22*, no. 2410 (1973)
361. Frieske, S.: Biul. Inst. Ochr. Roslin *45*, 19 (1969)
362. McNeal, F.H., Hodgson, J.M., Berg, M.A.: Can. J. Plant Sci. *49*, 155 (1969)
363. Patil, V.A., Kale, V.R.: Qualitas Plant. *2*, 187 (1975)
364. Schweizer, C.J., Ries, S.K.: Science *165*, 73 (1969)
365. Steenbjerg, F. et al.: Plant and Soil *36*, 475 (1972)
366. Vigne, J.T., Li, P.H.: Potato Res. *17*, 327 (1974)
367. Chavan, J.K., Ghonsikar, C.P.: Pesticides *12*, 25 (1978)
368. Wiedman, S.J., Appleby, A.P.: Weed Res. *12*, 65 (1972)
369. Fedtke, C.: $^{14}CO_2$-fixation by wheat plants treated with methabenzthiazuron. In: Proc. VIII. Int. Plant Prot. Congr., Section III, Part I, p. 232, Moscow 1975
370. Zweig, G., Ashton, F.M.: J. Exp. Bot. *13*, 5 (1962)
371. Sikka, H.C., Davis, D.E.: Weed Sci. *17*, 122 (1969)
372. Wu, M.T., Singh, B., Salunkhe, D.K.: Experientia *28*, 1002 (1972)
373. Decleire, M., de Cat, W., Bastin, R.: Biochem. Physiol. Pfl. *169*, 409 (1976)
374. Decleire, M., van Roey, G.: Biochem. Physiol. Pfl. *174*, 240 (1979)
375. Truelove, B., Jones, L.R., Davis, D.E.: Weed Sci. *21*, 24 (1973)
376. Ibenthal, W.-D., Heitefuss, R.: Phytopath. Z. *95*, 193 (1979)
377. Wilfert, U., Weltzien, H.C.: Z. PflKrankh. PflSchutz, Sonderh. IX, 219 (1981)
378. Heitefuss, R.: Z. PflKrankh. PflSchutz, Sonderh. VI, 79 (1972)
379. Horsfall, J.G., Dimond, A.E.: Z. PflKrankh. PflSchutz LXIV, 415 (1957)
380. Nadar, H.M., Clegg, M.D., Maranville, J.W.: Plant Physiol. *56*, 747 (1975)
381. Ebert, E., van Assche, C.J.: Experientia *25*, 758 (1969)
382. Copping, L.G., Davis, D.E.: Weed Abstr. *21*, no. 2034 (1972)
383. Jordan, J.S. et al.: Weeds *14*, 134 (1966)
384. Copping, L.G., Davis, D.E., Pillai, C.G.P.: Weed Sci. *20*, 274 (1972)
385. Pillai, C.G.P., Davis, D.E.: Weed Sci. *21*, 461 (1973)
386. Copping, L.G., Davis, D.E.: Weed Sci. *20*, 86 (1972)
387. Torigoe, Y. et al.: Phytochem. *11*, 1623 (1972)
388. Dekhuijzen, H.M., Overeem, J.C.: Phytochem. *11*, 1669 (1972)
389. Skoog, F., Armstrong, D.J.: Ann. Rev. Plant Physiol. *21*, 359 (1970)
390. Takahashi, S. et al.: Phytochem. *17*, 1201 (1978)
391. Poscar, et al.: Life Sci. *7*, 699 (1968)
392. Kefford, N.P., Bruce, M.J., Zwar, J.A.: Phytochem. *12*, 995 (1973)

IX. Herbicides Accepting Electrons from the Photosynthetic Pigment System I

1. Interference with Photosynthetic and Other Electron-Transport Systems

As may be derived from the heading of this chapter, the herbicidal compounds to be discussed here (Table 23) may undergo reversible reduction and reoxidation. The reduction of the herbicide molecule produces a stable radical which may be reoxidized by molecular oxygen. The latter, by accepting a single electron, is converted into the highly reactive superoxide anion (p. 60) which appears to be the actually phytotoxic species.

Table 23. Herbicidal and related compounds accepting electrons from the photosynthetic pigment system I

Common name	Structure	Water sol. ($g\,l^{-1}$)
Cyperquat	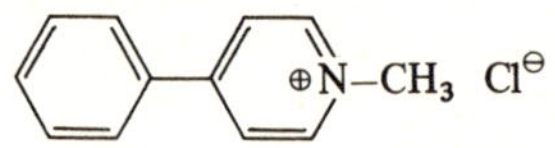	–
Difenzoquat	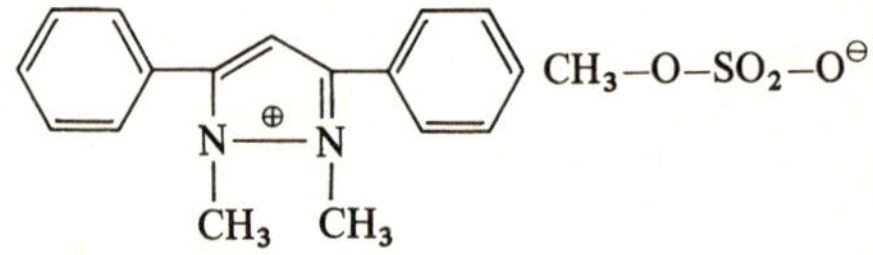	760
Diquat	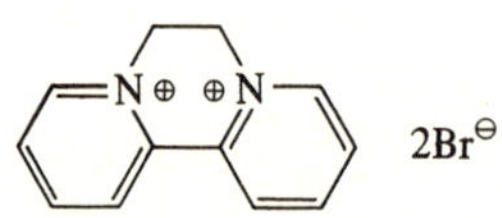	700
Metronidazole	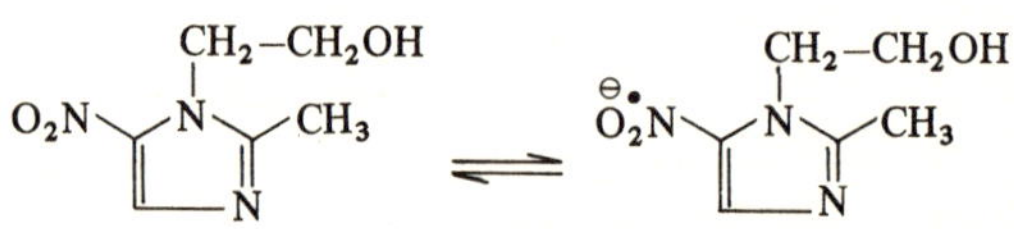	–
Morphamquat	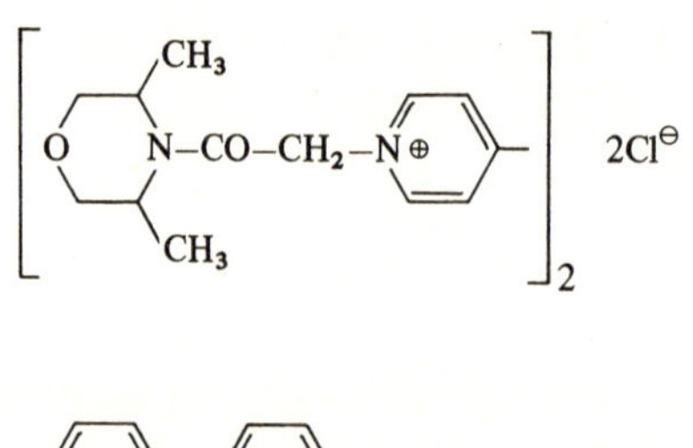	–
Paraquat	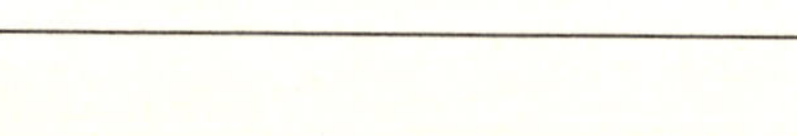	–

The most important biological system reducing the bipyridylium herbicides *diquat* and *paraquat* (also called methylviologen) and related compounds (Table 23) in the plant is the photosynthetic electron transport system. More precisely, the photosynthetic pigment system I mediates the transfer of low potential electrons (Fig. 14) to the herbicide molecule. Therefore, the electron transport system is not blocked as with the usual photosynthesis-inhibiting herbicides (p. 24), but the electrons are deviated from their normal path. Consequently, a pI_{50}-value cannot be calculated from in vitro measurements with isolated chloroplasts. Instead, a saturating concentration may be given which allows the maximum rate of pseudocyclic electron transport to be attained [1,2] (electron transport and photophosphorylation in the presence of an autoxidizable electron acceptor, which is the herbicide in this case; also called Mehler reaction) (Fig. 27 and p. 63). Half-saturating concentrations of herbicides are usually in the range of 0.1 µg/ml. However, photodestructions caused by *paraquat* and possibly also by related compounds may also occur in the absence of photosynthetic electron transport [17]. "Normal" *paraquat* interference with the photosynthetic electron transport is described by the following scheme:

$$2H_2O \xrightarrow{\text{PS II}} H_2O + 2H^+ + 2e^- + {}^1/_2O_2$$

$$2PQ^{2+} + 2e^- \xrightarrow{\text{PS I}} 2PQ^{\cdot+}$$

$$2PQ^{\cdot+} + 2O_2 \xrightarrow{\text{spontaneously}} 2PQ^{2+} + 2O_2^{\cdot-}$$

$$O_2^{\cdot-} + O_2^{\cdot-} \xrightarrow{\text{SOD}} O_2 + O_2^{2-}$$

$$O_2^{2-} + 2H^+ \xrightarrow{\text{spontaneously}} H_2O_2$$

$$\Sigma H_2O + {}^1/_2O_2 \longrightarrow H_2O_2$$

} oxygen uptake in the presence of KCN or NaN_3

In the presence of uninhibited catalase:

$$H_2O_2 \longrightarrow H_2O + {}^1/_2O_2$$

Σ Pseudocyclic electron flow

(PS I, II = photosystems I, II; PQ = paraquat; SOD = superoxide dismutase; catalase in chloroplast preparations is a cytoplasmic contaminant and may be inhibited by azide or cyanide)

The requirements for a molecule to mediate pseudocyclic electron transport and to be herbicidally active may be summarized as follows:

1. the redox potential should be between −300 and −714 mV [1,3],
2. a stable, water-soluble radical should be formed upon one-electron transfer [4–6],
3. the radical must be able to transfer its electron to molecular oxygen; very stable radicals which cannot react with O_2 are not herbicidally active [7].

Many compounds have been tested but only a few of them appear to be of practical use. For the bipyridylium herbicides *diquat* and *paraquat* a coplanar configuration is required for herbicidal activity [6]. Other active compounds include tetrazines [8], diquaternary salts of di-2-pyridyl ketone [9], and metronidazole [10–12]. Whereas fairly high concentrations (0.1–1.0 mM) are required for *metronidazole*, *cyperquat* [13], *difenzoquat* [14, 15], and *morphamquat* [16] are highly active commercial herbicides.

Under field conditions, the photosynthetic system is obviously the most important system to reduce the herbicide, thereby initiating damage in the tissue. This may easily be seen from the light dependence of the occurring damage. However, other biochemical processes might also activate these herbicides and generate damages, although much more slowly and at higher concentrations. Schneider et al. [18] reported that etiolated peas were inhibited by *diquat*, with the roots being most sensitive. *Paraquat* sprayed at a concentration of 1 mM equally initiates membrane leakage in green and in yellow soybean leaves in the light [19]. *Paraquat* also initiates limited breakdown of chloroplast membranes after 30 h in the dark [20]. *Paraquat* in 1 mM solution rapidly damages protoplast membranes [21] but exerts no direct effect on membrane integrity at 10 μM [22]. *Diquat* and *paraquat* also reveal little effect on red beet root cell membranes [23]. *Paraquat* at concentrations of 1 μM to 1 mM progressively inhibits DNA-synthesis and the completion of the cell cycle in *Vicia faba* root tips [24]. However, *paraquat* does not inhibit mitochondrial electron transport at 1 mM [25]. *Difenzoquat* exhibits its highest activity when applied to the leaf or plant base and is therefore presumed to act primarily on meristematic and young leaf tissue [26].

Admittedly, there is not very much evidence for an action of the "energy-deviating herbicides" on other than photosynthetic tissues. Still, some data clearly show that non-photosynthetic tissue may be sensitive to herbicidal action at high concentrations, especially after longer incubation times. The herbicide may therefore be reduced at a low rate by an electron transport system other than the photosynthetic electron transport system. In addition, at higher concentrations the ionic nature of the molecule may be the basis for a direct interference with membrane structure and integrity.

2. Physiology of Action and Selectivity

Besides the toxic radical effects to be described in the next chapter, inhibition of CO_2-fixation is the direct consequence of the occurring energy deviation. Alexander et al. [27] found that *paraquat* not only inhibits photosynthetic CO_2-fixation but also interferes with photosynthate translocation in the dark. Inhibition of photosynthetic CO_2-fixation has also been described to be a primary effect in protoplasts [22]. The ultrastructural damage caused by *paraquat* in *Pinus elliottii* cell suspension cultures has been described by Birchem et al. [28]. Destructive effects on vacuolar and organelle membranes comparable to normal cell senescence were observed although occurring more rapidly at concentrations between 1 and 10 $mg l^{-1}$. Oleoresin production, initiated by *paraquat* in intact *Pinus elliottii* tissue, has not been observed in the cell suspension cultures.

Since *diquat* and *paraquat* act very rapidly in full sunlight, very little translocation out of the treated leaves takes place under these conditions. *Diquat* and *paraquat* may therefore be applied as "desiccators". However, in the absence of photosynthetic electron transport and resulting rapid herbicidal action, these herbicides are translocated out of the treated leaves. Therefore, whereas in full sunlight toxicity occurs very rapidly within a few hours, at low light intensities or in the shade the herbicidal activity holds longer and the toxic symptoms are more evenly spread [29]. Similarly, the simultaneous application of photosynthesis-inhibiting herbicides *(bromacil, monuron, diuron)* synergistically increases the herbicidal activity and the duration of weed control by bipyridylium herbicides. Brian [30] studied the correlations between *paraquat* uptake, translocation and activity by applying different light regimes and light qualities; he concluded that also endogenous differences in tissue sensitivity contributed to the differences in herbicidal activity.

Harvey et al. [31, 32] and Harper et al. [33] investigated *paraquat* tolerant lines of *Lolium perenne*. They compared the activities of catalase, peroxidase and superoxide dismutase in tolerant and in sensitive lines and found the activities of these enzymes to be increased by 50–100% in the tolerant lines. No difference could be found for the uptake, transport and detoxification of *paraquat*. Since *paraquat* as well as the related herbicides are reduced by the photosynthetic pigment system I and are thought to initiate toxicity by transferring electrons to oxygen, producing $O_2^{\cdot}$ and H_2O_2, the increase of enzyme activities which destroy these toxic molecules in tolerant lines is in accordance with our present mode of action ideas.

An approach to the selection of resistant mutants has been suggested for *diquat* by Miles [34] and for *metronidazol* by Schmidt et al. [10]. The basic idea is that mutants defective in the photosynthetic electron transport are no longer killed by the herbicide because $O_2^{\cdot -}$ cannot be formed. By measuring the fluorescence induction it is then possible to differentiate between mutations before and within pigment system II (no fluorescence induction) and mutations after Q (fluorescence induction) as with photosystem II inhibitors (Fig. 14).

3. Sequence of Effects and Generation of Phytotoxicity

The requirement for both light and oxygen for the occurrence of phytotoxicity has repeatedly been demonstrated [35, 36]. As described previously (p. 60), the primary toxicant initiating the variety of destructive reactions is thought to be the superoxide anion, $O_2^{\cdot -}$. $O_2^{\cdot -}$ may directly oxidize unsaturated fatty acids and give rise to malondialdehyde (MDA) production (p. 68). In a detailed kinetic investigation, Harris and Dodge measured the changes in photosynthetic activities, pigment concentrations and membrane damage after *paraquat* application (Fig. 33). The process most rapidly inactivated was the photosynthetic CO_2-fixation. The electron flow from water to ferricyanide (Fecy) in the isolated chloroplast was inactivated much more slowly. The stability of the photosystem I exceeds that of the photosystem II [35, 38]. The increase in potassium efflux and the rise in the level of malondialdehyde occurred subsequent to the inactivation of the electron transport system. These two effects should be causally related since the occurrence of

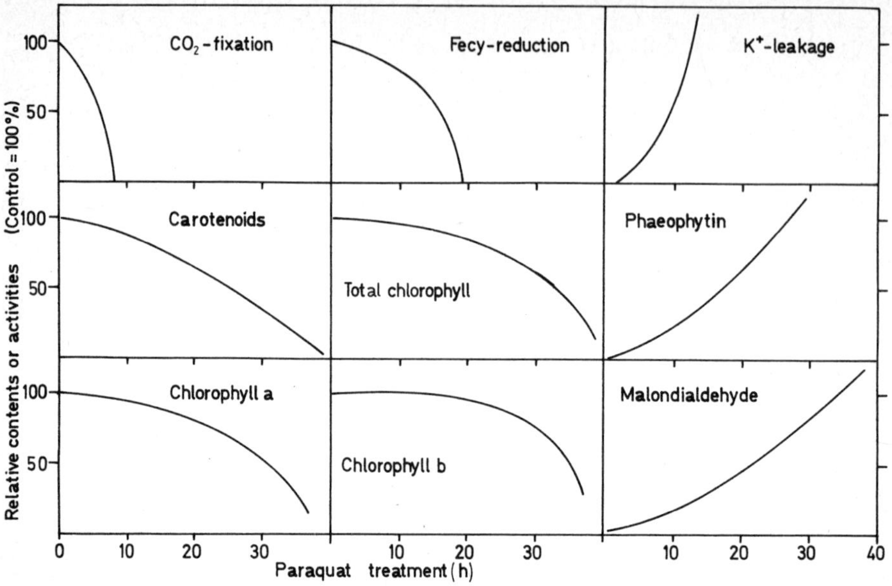

Fig. 33. Sequence of effects in flax cotyledon leaves after *paraquat* application. Phaeophytin and malondialdehyde given as percent increase with respect to control values, K^+-leakage in arbitrary units [37]

malondialdehyde indicates membrane damage. The destruction of carotenoids and chlorophylls bears much resemblance to the action of photosystem II inhibitors (p. 68), indicating a related mode of action including oxidative and radical damage, although occurring much more rapidly in the case of bipyridylium herbicides. The electron microscopic investigation parallelling the biochemical analysis revealed tonoplast breakdown after 6 h [20]. At later stages, a rapid and progressive deterioration of cell contents including the disruption of mitochondria and the breakdown of the chloroplast thylacoid membrane structure with the accumulation of osmiophilic globuli became evident. After *diquat* application a chloroplast swelling has already been observed after 3–4 h [16]. The further sequence of ultrastructural damage was as with *paraquat*.

The membrane damage may more directly be monitored by wide-angle X-ray diffraction measurements in chloroplast and other preparations. These measurements have shown that portions of the lipid bilayer become crystalline as the tissue senesces, as concluded from an increase in the transition temperature. *Paraquat* at low concentrations simulates the membrane phase properties of senescing tissue [39]. Conversely, these observations suggest that natural membrane senescence could be due to free radical damage. As mentioned before, high concentrations of *paraquat* directly destroy the membrane integrity [21].

Another consequence which may be drawn from the free-radical membrane damage induced by *paraquat* and related herbicides is the possibility of decreasing this damage through simultaneous application of free-radical scavengers. Accordingly, compounds with superoxide dismutase activity, e.g. copper chelates, effectively reduce paraquat-induced membrane damage [40, 41]. Giannopolitis et al.

found that *diuron* might also antagonize *paraquat* action by destabilizing O_2^- [42]. They generated O_2^- by means of xanthineoxidase or photochemically through riboflavin/methionine by reaction with p-nitro blue tetrazolium chloride. The ability of different herbicides to either produce or react with the superoxide anion was then examined. Whereas *paraquat* increased the availability of O_2^-, presumably by stabilizing it, *diuron* decreased O_2^-. This could in part contribute to the antagonism between *paraquat* and *diuron* in the initiation of rapidly occurring phytotoxicity after *paraquat* application.

A comprehensive survey of the bipyridylium herbicides has recently been published [43].

References for C. IX

1. Zweig, G., Hitt, J.E., Cho, D.H.: J. Agr. Food Chem. *17*, 176 (1969)
2. Halliwell, B.: Prog. Biophys. Molec. Biol. *33*, 1 (1978)
3. White, B.G.: Proc. 10th Brit. Crop Prot. Conf.-Weeds, p. 997 (1970)
4. Calderbank, A.: Chemical structure and biological activity of the bipyridinium herbicides. In: Herbicides, fungicides, formulation chemistry, Proc. II. Int. Congr. of Pestic. Chem. (IUPAC), Vol. V. Tahori, A.S. (ed.), p. 29. New York, London, Paris: Gordon and Breach 1972
5. Calderbank, A.: Adv. Pest Control Res. *8*, 127 (1968)
6. Baldwin, B.C.: On the mode of action of bipyridylium herbicides in photosynthesis. In: Progress in photosynthesis research, Vol. III. Metzner, H. (ed.), p. 1737. Tübingen: Laupp Press 1969
7. Summers, L.A., Black, A.L.: Nature *218*, 1067 (1968)
8. Pilgram, K.H. et al.: J. Agr. Food Chem. *25*, 888 (1977)
9. Black, A.L., Summer, L.A.: J. Chem. Soc. *17c*, 2394 (1970)
10. Schmidt, G.W., Matlin, K.S., Chua, N.-H.: Proc. Natl. Acad. Sci. U.S.A. *74*, 610 (1977)
11. Edwards, D.J. et al.: Z. Pflanzenphysiol. *71*, 424 (1974)
12. Tetley, R.M., Bishop, N.J.: Biochim. Biophys. Acta *546*, 43 (1979)
13. Schwarzbeck, R.A.: Proc. 12th Brit. Crop Prot. Conf. – Weeds, p. 851 (1974)
14. Winfield, R., Caldicott, J.B.: Pestic. Sci. *6*, 297 (1975)
15. Chow, P.N.P.: Can. J. Plant Sci. *56*, 429 (1976)
16. Dodge, A.D., Lawes, G.B.: Weed Res. *14*, 45 (1974)
17. Elstner, E.F., Lengfelder, E., Kwiatkowski, G.: Z. Naturforsch. *35c*, 303 (1980)
18. Schneider, I., Günther, G.: Biol. Zentralbl. *93*, 459 (1974)
19. Prendeville, G.N., Warren, G.F.: Weed Res. *30*, 251 (1977)
20. Harris, N., Dodge, A.D.: Planta *104*, 201 (1972)
21. Boulware, M.A., Camper, N.D.: Physiol. Plant. *26*, 313 (1972)
22. Davis, D.G., Shimabukuro, R.H.: Abstr. Weed Sci. Soc. Amer. Meet., p. 71 (1973)
23. Sutton, D.L., Foy, C.L.: Bot. Gaz. *132*, 299 (1971)
24. Bell, S.L., Schwarz, O.J., Hughes, K.W.: Can. J. Genet. Cytol. *18*, 93 (1976)
25. Foy, C.L., Penner, D.: Weeds *13*, 226 (1965)
26. Coupland, D., Taylor, W.A., Caseley, J.C.: Weed Res. *18*, 123 (1978)
27. Alexander, A.G., Biddulph, O.: Plant Cell Physiol. *17*, 601 (1976)
28. Birchem, R., Brown, C.L.: Stain Technol. *53*, 67 (1978)
29. Headford, D.W.R.: Pestic. Sci. *1*, 41 (1970)
30. Brian, R.C.: Pestic. Sci. *1*, 38 (1970)
31. Harvey, B.M.R., Muldoon, J., Harper, D.B.: Plant Cell Envir. *1*, 203 (1978)
32. Harvey, B.M.R., Fraser, T.W.: Plant Cell Envir. *3*, 107 (1980)
33. Harper, D.B., Harvey, B.M.R.: Plant Cell Envir. *1*, 211 (1978)
34. Miles, C.D.: Plant Physiol. *57*, 284 (1976)
35. Dodge, A.D., Harris, N.: Biochem. J. *118*, 43 (1970)
36. van Rensen, J.J.S.: Physiol. Plant. *33*, 42 (1975)
37. Harris, N., Dodge, A.D.: Planta *104*, 210 (1972)

38. Böger, P., Kunert, K.J.: Z. Naturforsch. *33c*, 688 (1978)
39. Thompson, J.E., Chia, L.S., Dumbroff, E.B.: Plant Physiol. Suppl. *63*, no. 414 (1979)
40. Youngman, R.J., Dodge, A.D.: Z. Naturforsch. *34c*, 1032 (1979)
41. Youngman, R.J. et al.: Experientia *35*, 1295 (1979)
42. Giannopolitis, G.N., Ries, S.K.: Weed Sci. *25*, 298 (1977)
43. Summers, L.A.: The bipyridinium herbicides, pp. 449. London, Toronto, Sydney, San Francisco: Academic Press 1980

X. Diphenyl Ether Herbicides

1. General Aspects

Diphenyl ether herbicides are characterized by diphenyl ether bridges which occur in all representatives, except in the recently introduced cyclic imides from which the experimentally applied compound *MK 616* is taken here as an example (Table 24). A second and more important reason for the formation of "diphenyl ether herbicides" is the dependence of their activity on light. Toxicity symptoms may appear within a few hours upon exposure to light of high intensity but become obvious in the dark only after several days or not at all. This light-dependence is, however, not absolute since some of the diphenyl ether herbicides – those having no orthosubstituent – are active both in the light and in the dark [1]. The latter group is commercially unimportant. Moreover, *nitrofen*, whose activity clearly depends on light [1, 2], may also be herbicidally active in the dark under certain conditions [3]. Other diphenyl ether herbicides, for which a light-dependence has been reported, include *acifluorfen-methyl* [4], *CNP* [1, 5], *fluorodifen* [6, 7], *MK 616* [8], *oxyfluorfen* [2, 9–11], and 3,5-dichloro-2-(4-nitrophenoxy) pyridine whose structure is similar to that of *nitrofen* except for the pyridine ring [12]. This pyridine analog also induces chlorosis. *Credazine*, which is also structurally dissimilar to other herbicides of this group, may exhibit a completely different mode of action because it produces growth regulator like effects [13, 14].

Diphenyl ether herbicides may then display at least two different modes of action, one requiring light and the other not. Virtually, nothing is known about the mode of action in the dark. The structural similarities between the different diphenyl ethers and the occasionally observed phytotoxicity of "normal" ortho-substituted diphenyl ether herbicides in the dark could, however, be an indication that related physiological processes are affected both in the light and in the dark. The possible nature of these processes will be discussed in the next chapter.

2. The Light-Dependent Herbicidal Action

As mentioned before, light is commonly required for the herbicidal action of the main diphenyl ether herbicides. The absorption of light energy required for this herbicidal action appears to be mediated by carotenoid pigments. Whereas green tissue and yellow mutants containing nearly normal amounts of carotenoids are sensitive, white mutants with very little or no carotenoids are resistant to herbicidal action. The white plants used were either carotenoid-free mutants or were free of carotenoids because they had been treated with a herbicide inhibiting the relevant

Table 24. Diphenyl ether and related herbicides. The pI_{50}-value relates to the inhibition of photosynthetic electron flow (cf. Table 4)

Common name	Structure	Water sol. (mg l^{-1})	pI_{50}
Azifluorfen-methyl	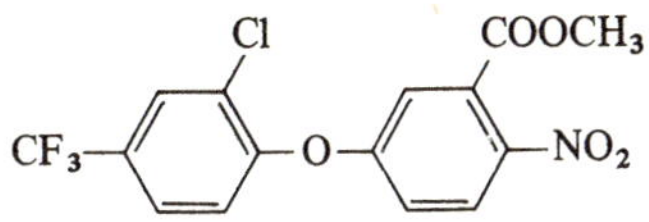	–	–
Bifenox	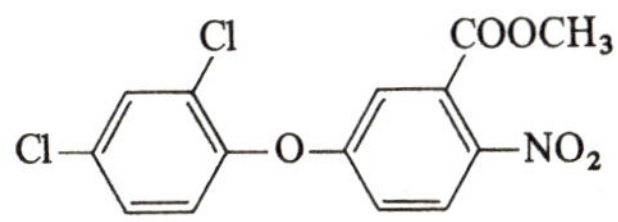	0.35	4.8
Chlomethoxynil	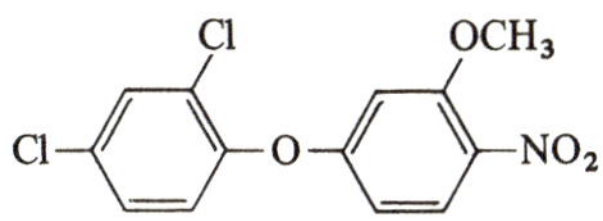	0.3	–
CNP	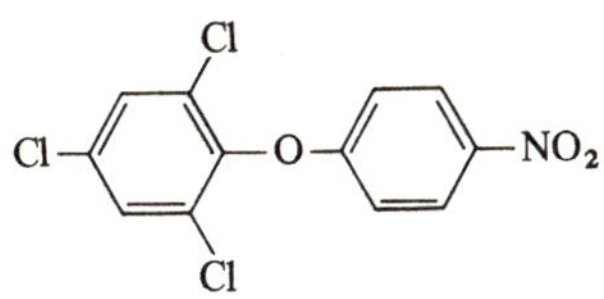	0.3	5.4
Credazine	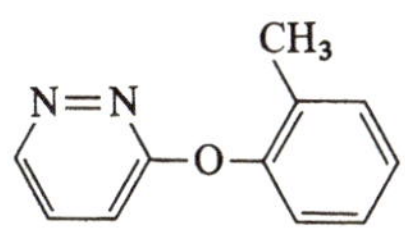	200	–
Fluorodifen	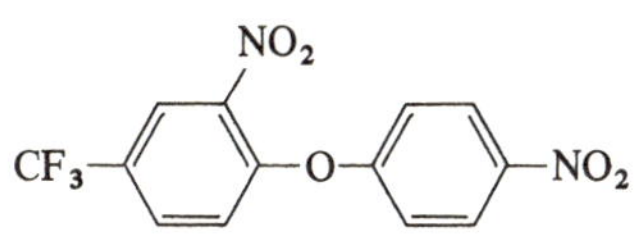	< 2	5.3
MK 616	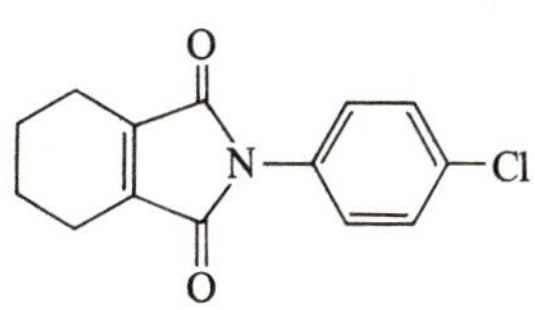	–	–
Nitrofen	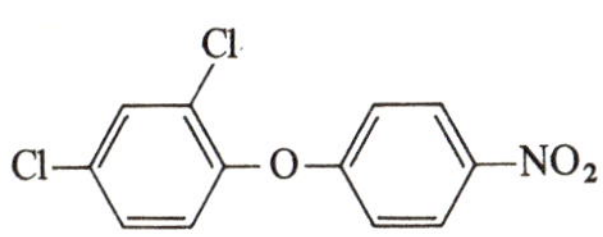	0.7–1.2	4.4
Oxyfluorfen	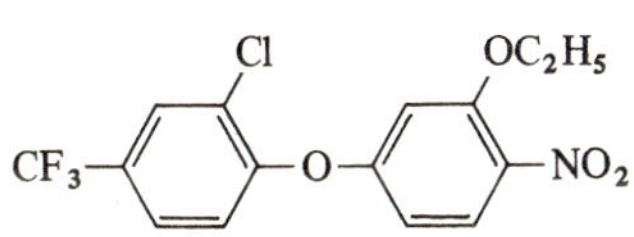	< 0.1	–

biosynthetic pathway (p. 99). This carotenoid requirement for herbicidal activity was first discovered for *nitrofen* by Matsunaka [1], and later confirmed for *oxyfluorfen* [2, 11], *azifluorfen-methyl* [4], *nitrofen* [2], and the cyclic imide herbicides [8]. Diphenyl ether herbicides without ortho-substituents, which are also active in the dark, act equally well on yellow and white mutants [1].

The requirement of absorption of light energy by carotenoid pigments, also called "light activation", for the generation of herbicidal activity by diphenyl ether herbicides has therefore clearly been established. "Light activation" may, however, not be an appropriate term because the herbicide needs not necessarily be "activated" by light but the acticity may simply depend on light as in the case of other compounds interfering with photosynthetic structures or processes. For *nitrofen* Matsunaka [1] has shown that "light activation" does not simply involve conversion into a toxic compound upon the action of light. There is also no indication that herbicide-sensitive plants catalyze the conversion of the parent compound into an active herbicide. For *oxyfluorfen* action in algae, a decrease of phytotoxicity was observed in the presence of *diuron* [15]. However, in similar applications of *azifluorfen-methyl* to higher plants, a similar interaction could not be observed [4]. Photosynthetic electron transport is therefore not necessary for phytotoxicity in higher plants. *Chlomethoxynil* is converted into the amino derivative and subsequently acetylated. The metabolites are further conjugated [19]. Tolerance for *fluorodifen* among cucumber (sens.) and peanuts (res.) is due to differences in herbicide uptake, transport, and metabolism [20–26]. Metabolism of *nitrofen*, on the other hand, was found to be very slow in herbicide-sensitive as well as in herbicide-resistant plants [27]. *Bifenox*, which was originally thought to be metabolized to *nitrofen*, mainly yields the free acid when metabolized in the soil [16]. Mobility differences have been found among plants either sensitive or resistant to bifenox [17]. Also *CNP* shows very little mobility and is metabolized only slowly [18]. In none of these cases was a metabolite with high herbicidal activity found.

Besides light and carotenoid pigments, oxygen is also required for herbicidal activity [4, 6, 15, 28]. This observation is particularly important because it relates the action of diphenyl ether herbicides to that of bipyridylium herbicides (p. 86) and inhibitors of photosystem II (p. 68). The important function of physiological and biochemical systems protecting against photooxidative damage has already been described (p. 60). The most important toxicants generated from ground-state triplet oxygen ($^3O_2$) are thought to be singlet oxygen ($^1O_2$) and the superoxide anion ($O_2^{\cdot-}$). Whereas singlet oxygen represents an oxygen molecule in an activated state, $O_2^{\cdot-}$ is formed by a one-electron transfer. Both types of activated oxygen may be formed after light excitation of the photosynthetic pigments (Figs. 27, 34). However, not only pigment excitation but also oxidative enzyme reactions may activate oxygen. Xanthine oxidase, aldehyde oxidase, and dihydro-orotic dehydrogenase produce $O_2^{\cdot-}$, as do several flavin dehydrogenases [29]. In microorganisms, but possibly also in higher cells, a considerable amount of the oxygen consumption results in $O_2^{\cdot-}$ production. Isolated mitochondria produce $O_2^{\cdot-}$. Consequently, systems either destroying $O_2^{\cdot-}$, like superoxide dismutase (SOD), or protecting otherwise against the toxicity caused by $O_2^{\cdot-}$, would therefore be interesting herbicidal targets. The same is true for $^1O_2$. $^1O_2$ is formed during spontaneous $O_2^{\cdot-}$ dismutation, but also by the photosynthetic pigment system.

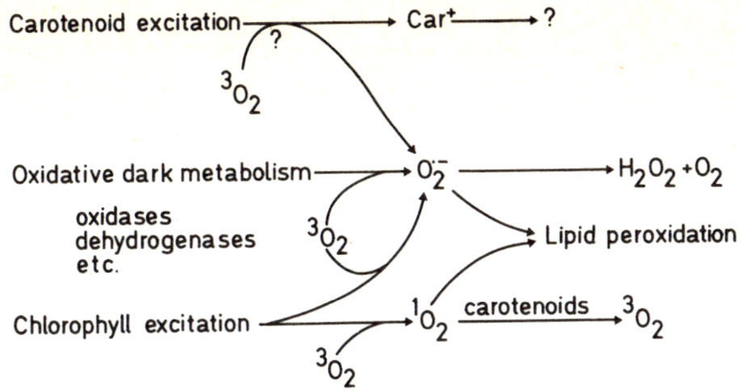

Fig. 34. Interrelationships of $^1O_2$ (singlet oxygen) and $O_2^{\cdot-}$ (superoxide anion) metabolism in leaf tissue

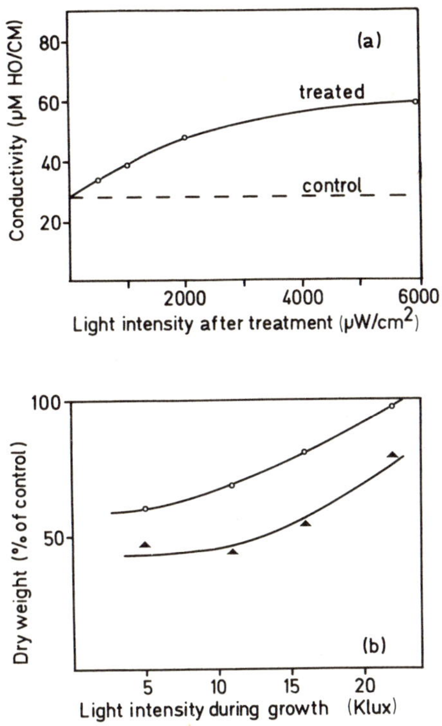

Fig. 35. Influence of the light intensity on **a** the toxicity of *oxyfluorfen* towards *Fagopyron esculentum* [10] and on **b** the growth rate of soybean and bean plants in the presence of *fluorodifen* [7]. Reproduced with permission of the Weed Science Society of America

It is not yet clear in which way carotenoids participate in the generation of phytotoxicity. The main carotenoids present in the sensitive yellow mutants are xanthophylls [1, 4]. Apparently, some photosensitization process is involved which activates oxygen. This activation can possibly occur in the presence as well as in the absence of a diphenyl ether herbicide, and the question of the precise interference of the herbicide remains unsolved. However, since $^1O_2$ is quenched rather than produced by the carotenoid molecule, the product of carotenoid photosensitization might again be $O_2^{\cdot-}$ [30]. This interpretation, however, does not take into account the origin of the necessary reducing equivalents.

Although it may be assumed that diphenyl ether herbicides interfere with some systems protecting against $O_2^{\cdot-}$ induced damage, there is only indirect evidence available supporting this hypothesis: diphenyl ether herbicides act rapidly and, in the case of carotenoid-sensitization, the activity depends on light and on molecular oxygen; the lipid peroxidation product malondialdehyde which indicates oxidative membrane damage is produced during herbicidal action [4, 15, 28]. Moreover, the contradictory reports on the activity of different diphenyl ether herbicides in the light or in the dark may indicate that analogous protective systems may be inhibited in either case.

Not only may *nitrofen* be phytotoxic in the dark [3], but also the effect of light intensity may differ. Whereas *oxyfluorfen* became increasingly active with increasing light intensity (Fig. 35a), *fluorodifen* in another experiment was more toxic at lower light intensities (Fig. 35b). However, no measurements were performed below 5,000 lx. The different behaviour in the experiments recorded in Figs. 35a, b may be the consequence of different treatment schedules. It may, however, also reflect a different physiological status of the various treated plants if we assume that also dark reactions may constitute a basis for herbicidal interference.

Since red and yellow light were more active than blue and white light, also chlorophyll molecules may be involved in the normal photosensitization process in green tissue.

The herbicidal activity resulting from diphenyl ether herbicide application may be increased by treatment in the night or by artificially creating darkness for a few hours after treatment [2, 31]. The usually accepted interpretation assumes a better and more uniform distribution of diphenyl ethers within the treated plant, as with bipyridylium herbicides (p. 89).

3. Influence of Diphenyl Ether Herbicides on Photosynthesis and Respiration

The requirements for the herbicidal action of diphenyl ether herbicides differ remarkably from those of other light-dependent herbicides. Diphenyl ether herbicides act very rapidly in full sunlight (within a few hours). This activity does not require the presence of chlorophyll but only that of xanthophylls.

A primary inhibition of photosynthetic or respiratory processes should therefore not be the main cause of the damage which eventually occurs in the tissue. Nevertheless, inhibitory and uncoupling actions of diphenyl ether herbicides on mitochondrial and chloroplast electron transport processes have been found and may play an additional role in herbicidal action, preferrably at high concentrations.

The influence of five diphenyl ether herbicides on electron transport in chloroplasts and mitochondria is shown in Table 25. Although the herbicide concentrations required to cause 50% inhibition of photosynthetic electron transport are somewhat higher than in the case of "normal" photosystem II inhibitors (roughly one order of magnitude), they would not exclude the possibility of in vivo-photosynthesis inhibition. However, no inhibition of CO_2-fixation could be found

Table 25. Inhibition by diphenyl ether herbicides of electron transport in isolated chloroplasts and mitochondria [32, 33, 42]

Common name	I_{50}(M)	
	Chloroplast electron transport	Mitochondria state 3 respiratory O_2 uptake
Bifenox	1.5×10^{-5}	2.3×10^{-5}
CNP	4.0×10^{-6}	1.6×10^{-5}
Fluorodifen	5.6×10^{-6}	$>10^{-4}$
Nitrofen	3.3×10^{-5}	2.7×10^{-5}
Oxyfluorfen	10^{-4}	–

in the early course of herbicidal action of *CNP* [5] and *oxyfluorfen* [10]. On the contrary, interference with photosynthetic phosphorylation was repeatedly suggested as a herbicidal mechanism. However, the inhibition of photophosphorylation in isolated chloroplasts is not very strong: about 10 μM *nitrofen* [3, 34] are required to effect 50% inhibition. From experiments with the alga *Bumilleriopsis* Lambert et al. [34] classified *nitrofen* as an energy-transfer inhibitor. Whereas this may be true for the intact algae system, the relevance for the intact plant is not clear. Luciferase measurements of the ATP level in diphenyl ether herbicide-treated bean leaves did not show a significant decrease at up to 50% tissue damage in the light [35]. In contrast to the photosystem II inhibitors, which uncouple photosynthetic phosphorylation only at higher concentrations (Fig. 18), diphenyl ethers did the same at concentrations somewhat lower than those required for the inhibition of electron transport [32]. Research on the diphenyl ether inhibition site in the chloroplast electron transport has been reported [37]. Trebst [37] found that some diphenylether compounds interfere with the chloroplast electron transport system at the "DBMIB-site" (see p. 24).

In chloroplasts isolated from *CNP*-treated plants and tested for their electron transport activities, a decreased photophosphorylation rate was found [38]. Using the same preparations, the activity of photosystem I was increased whereas that of photosystem II remained unchanged. However, cyclic photophosphorylation was found to be inhibited in these chloroplasts, in contrast to the results of *nitrofen* action on isolated chloroplasts in which cyclic photophosphorylation was much less sensitive than non-cyclic photophosphorylation. It may therefore be concluded that photophosphorylation is the most sensitive system in vivo which is damaged first in the course of herbicidal action. For *fluorodifen* accumulation of the herbicide in the chloroplast/nuclei fraction of tomato protoplasts has been reported [39]. No effect on the integrity of protoplasts could be observed [40].

The concentrations required for the inhibition of mitochondrial electron transport are even higher than those required for the inhibition of chloroplast electron transport (Table 25).

State 3 respiration is measured at saturating ADP concentrations and represents the "uncontroled" maximum rate of oxygen uptake (p. 116). In contrast, *fluorodifen* and *nitrofen* were found to increase state 4 respiration at concentrations between 10 and 100 μM, which indicates uncoupling of respiratory energy conservation at these concentrations.

4. Credazine

Since the structure of *credazine* is similar to that of diphenyl ether herbicides it was included into the latter group of herbicides. Its mode of action appears, however, to be quite different. It has been assumed that only sensitive plants metabolize *credazine* into the actual herbicide, which might be o-cresol [41]. This, however, is improbable because o-cresol is not likely to cause regulatory growth effects which are observed after *credazine* application [13, 14]. *Credazine* inhibits cell division, cell elongation and growth but not photosynthesis, respiration, and germination, nor does it decrease the auxin level. The inhibition of growth and the morphological effects may be antagonized by simultaneous application of *IAA*. An interference with the action of auxin has therefore been suggested as the mode of action.

References for C. X

1. Matsunaka, S.: J. Agr. Food Chem. *17*, 171 (1969)
2. Fadayomi, O., Warren, G.F.: Weed Sci. *24*, 598 (1976)
3. Pereira, J.F., Splitstoesser, W.E., Hopen, H.J.: Weed Sci. *19*, 662 (1971)
4. Hess, F.D., Orr, G.L.: Abstr. Weed Sci. Soc. Amer. Meet., nos. 246–249 (1981)
5. Hyakutake, H., Ishizuka, K.: Selective action of 2,4,6-trichloro-4-nitro-diphenylether (CNP) on rice and barnyardgrass. In: IV. Int. Congr. of Pestic. Chem. (IUPAC), Abstr. Vol., no. IV-653 (1978)
6. van Assche, C.J., Ebert, E.: Sur le mode d'action du p-nitophenyl, α,α,α-trifluoro-2-nitro-p-tolyl-éther sur les vegetaux supérieurs. In: Compte Rendu des Journées d'Etudes sur les Herbicides, Vol. I, 6ᵉ Conference du Columa, p. 15 (1971)
7. Pollak, T., Crabtree, G.: Weed Sci. *24*, 571 (1976)
8. Wakabayashi, K. et al.: Mechanism of action of cyclic imide type herbicides. In: IV. Int. Congr. of Pestic. Chem. (IUPAC), Abstr. Vol., no. IV-655 (1978)
9. Vanstone, D.E., Stobbe, E.H.: Weed Sci. *25*, 352 (1977)
10. Vanstone, D.E., Stobbe, E.H.: Weed Sci. *27*, 88 (1979)
11. Prendeville, G.N., Warren, G.F.: Weed Res. *30*, 251 (1977)
12. Fujikawa, K. et al.: Agr. Biol. Chem. *34*, 68 (1970)
13. Takeuchi, Y., Konnai, M., Takematsu, T.: Weed Abstr. *23*, no. 1341 (1974)
14. Takematsu, T. et al.: 3-Phenoxypyridazine herbicide. In: Proc. fifth Asian-Pacific Weed Sci. Soc. Conf., p. 92 (1975)
15. Lambert, R., Böger, P.: Z. PflKrankh. PflSchutz, Sonderh. IX, 147 (1981)
16. Leather, G.R., Foy, C.L.: Pestic. Biochem. Physiol. *7*, 437 (1977)
17. Leather, G.R., Foy, C.L.: Weed Sci. *26*, 76 (1978)
18. Shimotori, H., Kuwatsuka, S.: J. Pestic. Sci. *3*, 267 (1978)
19. Niki, Y., Kuwatsuka, S., Yokomichi, D.: Agr. Biol. Chem. *40*, 683 (1976)
20. Eastin, E.F.: Weed Res. *11*, 63 (1971)
21. Eastin, E.F.: Weed Sci. *19*, 261 (1971)
22. Eastin, E.F.: Weed Res. *12*, 75 (1972)
23. Eastin, E.F.: Weed Sci. *20*, 255 (1972)
24. Eastin, E.F.: Weed Sci. *11*, 120 (1971)
25. Shimabukuro, R.H. et al.: Pestic. Biochem. Physiol. *3*, 483 (1973)
26. Frear, D.S., Swanson, H.R.: Pestic. Biochem. Physiol. *3*, 473 (1973)
27. Shimabukuro, R.H., Walsh, W.C.: Abstr. Weed Sci. Soc. Amer. Meet., no. 212 (1979)
28. Kunert, K.J., Böger, P.: Weed Sci. *29*, 169 (1981)
29. Fridovich, I.: Science *201*, 875 (1978)
30. Feitelson, J.: The role of oxygen in photosensitized reactions of proteins and nucleic acids. In: Research in photobiology. Castellani, A. (ed.), p. 235 (1976)
31. William, R.D., Warren, G.F.: Weed Res. *15*, 285 (1975)

32. Moreland, D.E. et al.: Weed Sci. *18*, 636 (1970)
33. Kruger, P.J. et al.: Weed Abstr. *24*, no. 2798 (1975)
34. Lambert, R., Kunert, K.-J., Böger, P.: Pestic. Biochem. Physiol. *11*, 267 (1979)
35. Fedtke, C.: unpublished
36. Bugg, M.W. et al.: Plant Physiol. *65*, 47 (1980)
37. Trebst, A., Draber, W.: Structure activity correlations of recent herbicides in photosynthetic reactions. In: Advances in pesticide science, Part 2. Geissbühler, H. (ed.), p. 223. Oxford, New York: Pergamon Press 1979
38. Hyakutake, H., Ishizuka, K.: Selective action of 2,4,6-trichloro-4-nitro-diphenylether (CNP) on rice and barnyardgrass. In: Proc. fifth Asian Pacific Weed Sci. Soc. Conf., p. 192 (1975)
39. Boulware, M.A., Camper, N.D.: Weed Sci. *21*, 145 (1973)
40. Boulware, M.A., Camper, N.D.: Physiol. Plant. *26*, 313 (1972)
41. Nakagawa, M., Kawakubo, K.: Agr. Biol. Chem. *35*, 764 (1971)
42. Pritchard, M.K., Warren, G.F., Dilley, R.A.: Weed Sci. *28*, 640 (1980)

XI. Herbicides Interfering with Carotenoid Biosynthesis

1. Herbicidal Structures Interfering with Carotenoid Biosynthesis

The herbicides and compounds which have become known to inhibit normal chloroplast development in the light, presumably by interfering either directly or indirectly with the biosynthesis of carotenoids, have very diverse structures (Table 26). Since some of these compounds exhibit similar structures they may be grouped together (e.g. pyridazinones, halogenated pyridines and related compounds) but a structural element common to all of them may not easily be discerned. Nevertheless, all compounds listed in Table 26 form chlorotic tissue. However, only the new growth appearing after treatment becomes chlorotic whereas preexisting green tissue mostly remains unaffected. The same is true for the greening of etiolated tissue (p. 104). These observations can satisfactorily be explained by the inhibition of carotenoid biosynthesis. However, it is not known whether the different structures all act at the same site. Moreover, at least for some of the herbicides classified into this group it is clear that more than one site is affected (pp. 107, 108).

Many herbicides are occasionally classified as "bleaching herbicides" because white tissue has been observed after treatment. However, herbicidal mechanisms other than the inhibition of carotenoid biosynthesis considered here lead to bleaching of preexisting tissue (diphenyl ethers, p. 92; bipyridyliums, p. 86). In a survey article Wolf [1] listed all compounds reported to date as "chlorosis inducers". Most of these compounds are not primarily inhibiting carotenoid biosynthesis; he also included into this group photosynthetic inhibitors of photosystem II which occasionally cause chlorosis in preexisting green tissue in very intense light (p. 64).

2. Physiological and Biochemical Changes in Treated Plants

a. Carotenoid Biosynthesis

In the following the pathway of carotenoid biosynthesis is first briefly summarized. Carotenoids are a family of compounds composed of the isoprenoid building blocks isopentenyl pyrophosphate and dimethylallyl pyrophosphate which, in turn, are formed from mevalonic acid [2]:

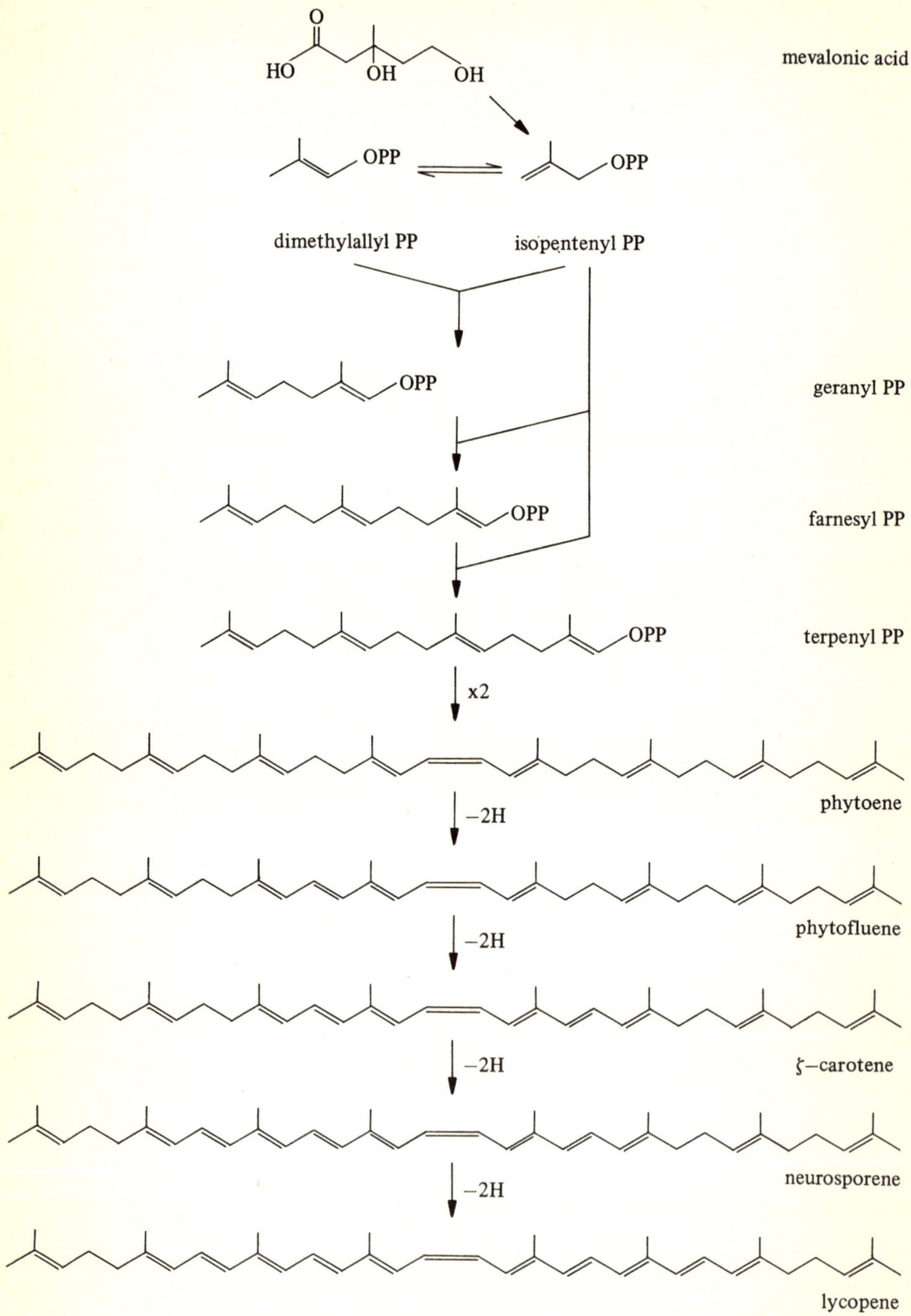

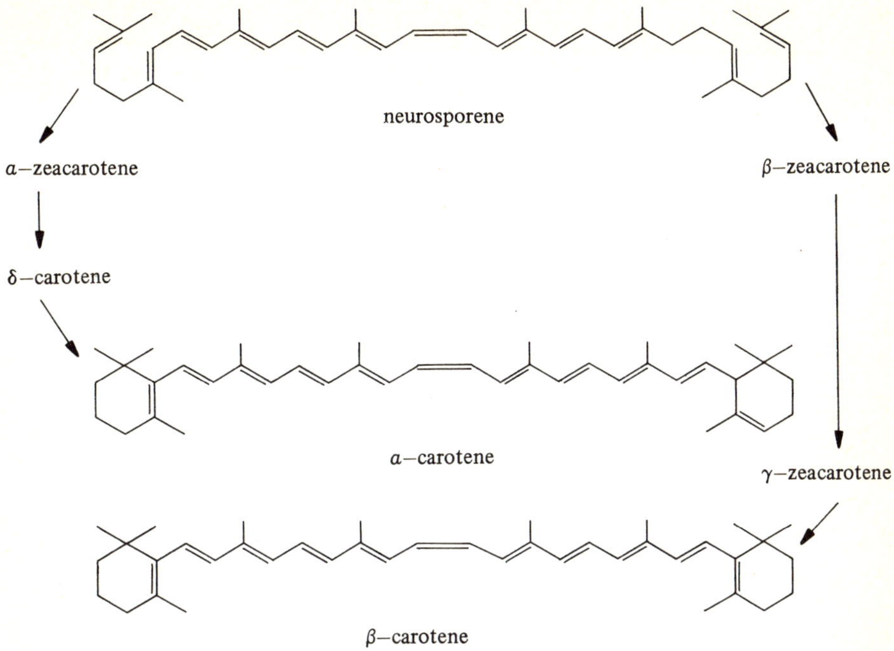

The synthesis includes three steps: (a) elongation in which four C-5 units are condensed consecutively to yield terpenyl pyrophosphate (geranylgeraniol pyrophosphate), two of these C-20 compounds being then condensed to yield phytoene, (b) dehydrogenation finally affording lycopene, the red pigment in tomatoes, and (c) cyclization eventually yielding different cyclized carotenes, starting from neurosporene. Carotenoid biosynthesis not only occurs in the cytoplasm, chloroplast, and etioplast of higher plants but also in many microorganisms.

b. Carotenoids in Herbicide-Treated Plants

The principal carotenoid precursors that accumulate in herbicide-treated tissue are phytoene, phytofluene, and ζ-carotene. They have been observed after treatment with *aminotriazol* [3–5], *dichlormate* [3,4], *difunon* [6,7], *fluridone* [8], *metflurazone* [9–13], *norflurazon* [8,14], *NK 049* [15], and *pyrichlor* [3,4]. This implies that basically dehydrogenation reactions are inhibited, but if some dehydrogenation does occur, the partially dehydrogenated intermediates are not cyclized. A few reports on temporary β-carotene formation in the presence of *aminotriazol* [5,13], *difunon* [7], and *metflurazone* [10,16] may be explained by limited transport of the herbicide.

It is still unclear whether all the herbicides inhibiting carotenoid biosynthesis interfere with the same inhibition site. On the contrary, it seems unlikely that the presently known diverse structures (Table 26) inhibit the complex biosynthetic pathway at the same site. *Aminotriazole*, *pyrichlor*, and *dichlormate* which form one group of herbicide, lead to the accumulation of phytoene, phytofluene, ζ-carotene,

Table 26. Herbicides inhibiting carotenoid biosynthesis

Common name	Structure	Water sol. ($mg\ l^{-1}$)
Aminotriazole	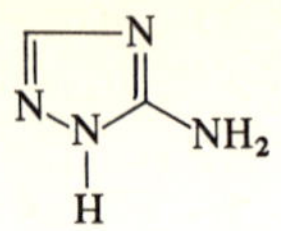	(28%)
Dichlormate	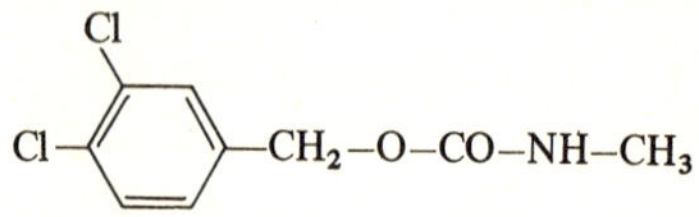	170
Difunon	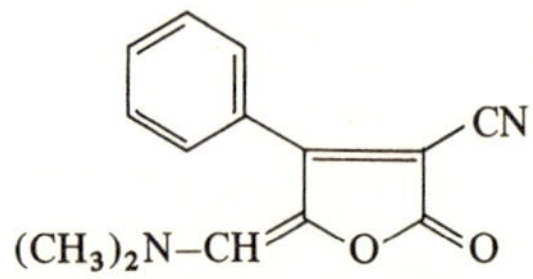	–
Fluometuron	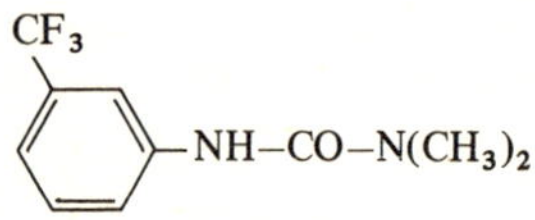	80
Fluridone	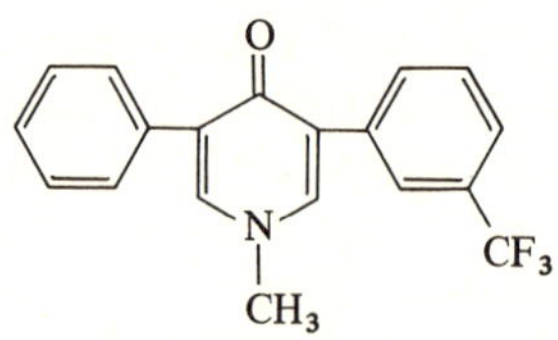	12
Haloxydine	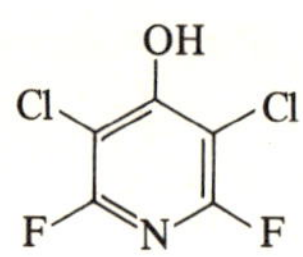	–
Ioxysone, Bromoxysone Hal = J, Hal = Br	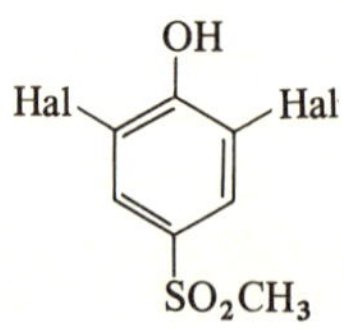	–
Metflurazone	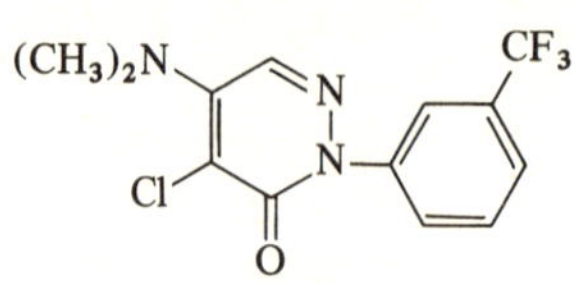	
Norflurazon	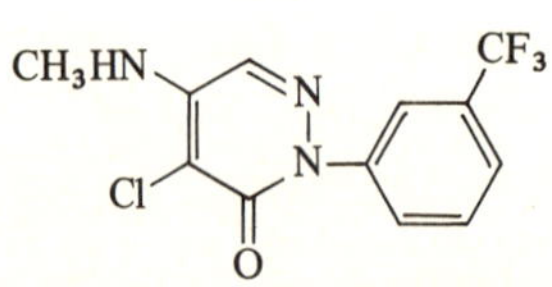	28

Table 26 (continued)

Common name	Structure	Water sol. (mg l^{-1})
NK 049 = Methoxyphenone	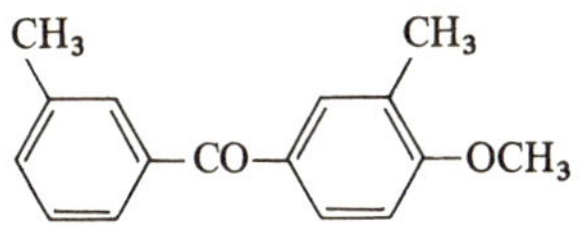	1.8
Pyriclor	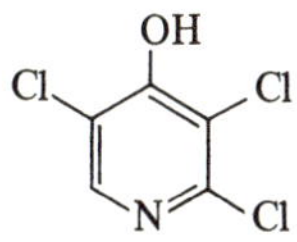	570
U 18	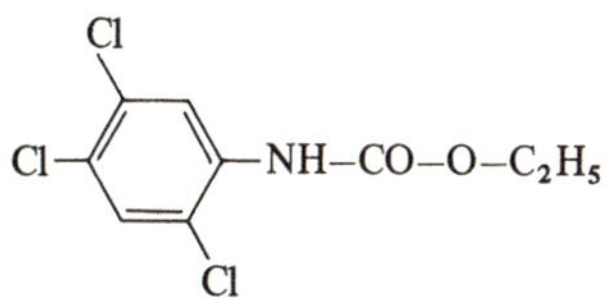	–

and even lycopene [13]. The other inhibitors of carotenoid synthesis mentioned almost exclusively give rise to the accumulation of phytoene, this generally leading to the assumption that the dehydrogenation reactions are primarily inhibited [6, 8, 14, 17]. For *metflurazone* and *norflurazon*, inhibition of the carotenoid biosynthetic pathway was confirmed in cell-free systems [17a, 18a]. Inhibition of dehydrogenation reactions also appears to be the mechanism of carotenoid biosynthesis inhibition by *metflurazone* in *Myxococcus flavus* [18]. In addition, the cyclization may directly or indirectly be inhibited under certain conditions, but other possible inhibition sites in the isoprenoid pathway are also discussed [10, 13]. A specific inhibitor of lycopene cyclization has been reported [19].

To sum up, inhibition of dehydrogenation reactions in the biosynthetic pathway leading to coloured carotenoid pigments appears to be the main biochemical mode of action for the herbicides listed in Table 26, with the exception of *haloxydine* [20, 21]. However, it is not known whether this is a direct or an indirect inhibition, nor is it known whether all of these compounds interfere with the same inhibition site.

c. The Influence of Light

In plants grown in the dark in the presence of one of the herbicides listed in Table 26 the common basic effect that can be observed is an inhibition of carotenoid biosynthesis. In the light a more complicated situation arises because secondary effects must be considered. The influence of light is most clearly demonstrated by the type of experiment presented in Table 27. Chlorotic tissue may only be ob-

Table 27. Conditions for obtaining chlorotic tissue after treatment with herbicides that inhibit carotenoid biosynthesis

Germination and early seedling growth		Plant Status	After seedling development		Plant Status
Light intensity (lx)	Herbicide		Light intensity (lx)	Herbicide	
0	None	Etiolated, yellow	≥10,000	None	Normal greening
0	None	Etiolated, yellow	≥10,000	Present	Normal greening
0	Present	Etiolated white	≥10,000	Present	Chlorotic
10–500	None	Green	10–500	None	Green
10–500	None	Green	10–500	Present	Green
10–500	Present	Green	10–500	Present	Green
10–500	None	Green	≥10,000	None	Green
10–500	None	Green	≥10,000	Present	Green
10–500	Present	Green	≥10,000	Present	Chlorotic

tained when the herbicide has been applied before germination and the young plant is irradiated with intense light (in excess of about 10,000 lx). Very low intensity irradiation yields pale green tissue.

The basis for an understanding of this behaviour lies in the role the carotenoids play in the protection of chlorophyll against photooxidation (p. 63): (a) in the dark no chlorophyll is formed in control and in treated plants; the plants etiolate in either case, whereas in the presence of a herbicide they are completely white; (b) in very low light intensities (10–20 lx) normal greening and chloroplast development occur which means that carotenoids are not essential for the formation of the chloroplast lamellar system; (c) with increasing light intensities the chlorophyll content of the treated plants progressively decreases, and in the presence of several thousand or more lx the treated plants are white. Also, herbicide-treated plants which have become green at 10–20 lx are bleached completely in intense light, indicating the photooxidative destruction of chlorophyll pigments in the absence of carotenoids. However, when the herbicide is applied after the first stages of seedling development, allowing for a normal carotenoid biosynthesis, etiolated tissue normally turns green. Only newly developed leaves are then again chlorotic.

Herbicides for which chlorophyll formation at very low light intensities and bleaching at higher light intensities has been demonstrated include *aminotriazole* [3], *pyrichlor* [3], *fluridone* [8,22], *metflurazone* [9,23], and *norflurazon* [8,24]. *Haloxydine* appears to act also on preexisting, normal green tissue, although only in the light [20].

Chlorophyll bleaching is only the most obvious and clearly visible phenomenon in plants treated with herbicides of this group. The photooxidative damage extends further to the destruction of chloroplast DNA [23], fraction I protein (ribulose-

1,5bisphosphate carboxylase) [25], 70 S ribosomes [22, 25], and also of the chloroplast endolamellar system [25, 26]. Favali et al. [27], who also reported on the lack of 70 S ribosomes, fraction I protein, and stroma lamellae in plants treated pre-emergence with *aminotriazole*, observed lamellar decay together with the formation of "super grana" after post-emergence treatment. They compared their results with those from mutant plants with low content of carotenoids.

d. The Generation of Phytotoxicity

As has been described above, the bleaching of herbicide-treated plant tissue is only obtained at higher light intensities. In intense light, excess excitation energy of chlorophyll pigments may be destroyed by one out of several pathways (p. 63). One of the most important protective mechanisms is the quenching of activated singlet oxygen by β-carotene (Figs. 27, 28). Singlet oxygen ($^1O_2$) is inevitably produced during photosynthesis in an oxygen-containing atmosphere. Under normal photosynthesizing conditions, $^1O_2$ reacts at a high rate with β-carotene and is converted into its triplet state. The ^{3}carotene is decomposed spontaneously, emitting the surplus energy as heat. β-Carotene may also quench the activated triplet chlorophyll directly by triplet-triplet resonance energy transfer [92]. To sum up, the main function of carotenoids in the photosynthetic pigment system appears to be the protection against photooxydative damage [28]. A very small portion of the β-carotene present appears, however, to be bound to the reaction centre and may function in primary photochemistry [29, 92].

In the absence of carotenoids, or at highly reduced carotenoid levels, the protective mechanism described above can obviously no longer be effective. The consequences, in addition to those described above, also include lipid peroxidation initiated by singlet oxygen (p. 60) [30, 31] and resulting in the formation of e.g. malondialdehyde (p. 68) [32].

In the sequence of effects that may occur after the application of herbicides that inhibit carotenoid biosynthesis, inhibition of photosynthetic CO_2-fixation and/or O_2-evolution, as has been reported after application of *aminotriazole* [33, 34] and *difunon* [32], must be assumed. However, no direct inhibition of photosynthetic processes could be observed upon application of these herbicides. Interestingly, accumulation of peroxidation products or of phytol, which would have to be expected if the synthesis of chlorophyll and of other chloroplast constituents proceeded in the presence of the herbicide and thereby supplied substrates for continuous peroxidations, usually does not occur [12, 13]. Therefore, a correlative synthesis of the different thylacoid-membrane constituents, including pigments, lipids, etc., should be taken into account.

When trying to understand the sequence of effects occurring between the primary herbicidal block and the final appearence of phytotoxicity, two possible types of applications should be considered separately: pre-emergence/seed treatment and post-emergence. After seed treatment, the situation is less complicated in that obviously the correlative synthesis of different thylacoid membrane materials avoids the accumulation of excess precursors, or of degradation products. However, the details are not clear because some photooxidation does occur, as evidenced by experiments using light of low intensity (Table 27). The further synthesis of chloroplast material in irradiated chloroplasts lacking carotenoids may be stopped at an early stage of development.

After post-emergence treatment of preexisting green tissue the situation is different. Some herbicides including *aminotriazole*, *metflurazone*, and *norflurazon* exhibit no immediate effect at all. *Difunon* leads to carotenoid breakdown in the unicellular green alga *Scenedesmus acutus* [32]. This, however, has only been observed in the air; under a nitrogen atmosphere very little carotenoid decomposition was found. *Haloxydine* [20] and *pyrichlor* [36] even induce swelling and breakdown of fully developed chloroplasts in the light.

A turnover of carotenoids does occur in developed green leaves, and in the absence of light the mature chloroplast degenerates into an etioplast or a leucoplast. However, the normal rate of carotenoid turnover does not explain this rapid light-dependent action on developed green plants. We must therefore consider other or additional sites of action for at least some of the herbicides simply classified as "inhibitors of carotenoid biosynthesis".

e. Microbody Relations

The leaf microbody or peroxisome contributes significantly to the protection of the leaf against photooxidative damage. This organelle is the site of the photorespiratory pathway which some authors regard as a system protecting against destructive photooxidations. Specifically, this organelle is the site of high catalase activity which is usually correlated with the chlorophyll content [37]. The contents of the peroxisomal enzymes catalase, glycolate oxidase, and hydroxypyruvate reductase increase considerably during the greening of etiolated tissue [38, 39]. Interestingly, *aminotriazole* inhibits the formation of catalase but not of the other two enzymes [38–40], and so did also *haloxydine* [41].

On the other hand, *difunon*, *fluometuron*, and *metflurazone* strongly decrease the activities of these three peroxisomal enzymes [41]. Also with respect to carotene formation and subsequent greening in etiolated seedlings, *aminotriazole* and *haloxydine*, which are not as rapidly active as the other relevant herbicides, form an own group [21]. This evidence is inconsistent with the idea of classifying all these compounds as "inhibitors of carotenoid biosynthesis". Clearly, they do not all behave similarly with respect to peroxisomal marker enzymes [42]. In view of the role the peroxisomes/leaf microbodies play in the protection against photooxidations, they may even be considered as a target for some of the chlorosis-inducing herbicides. Alternatively, "the primary effect of some of these herbicides may relate to some more basic biosynthetic or differentiating step needed for the light-stimulated biosynthesis of both the chloroplasts and the leaf microbodies" [41].

However, a completely different view may also be valid: since the normal peroxisome development takes place only after its attachment to the outer chloroplast membrane, the effects observed in peroxisomes may altogether occur secondarily.

3. Special Aspects of Individual Herbicides Inhibiting Carotenoid Biosynthesis

a. Aminotriazole

Early experiments with *aminotriazole* revealed an interference of this herbicide with amino acid and purine metabolism [43]. Specifically, imidazolylglyceryl phosphate and imidazolyl glycerol were found to accumulate in the treated plant, and this accumulation could be circumvented by the addition of 1-histidine, which provides evidence for an in vivo end product inhibition [44,45]. Besides the implicated inhibitory site in the histidine biosynthetic pathway, probably the dehydrogenase enzyme, other sensitive enzymatic steps have been discussed in the biosynthesis of purines and in the C-1 metabolism [4]. In the bacterium *Escherichia coli* growth inhibition by *aminotriazole* could be prevented by the addition of adenine. The *aminotriazole* concentrations used in these experiments were usually in the 0.1–1.0 mM range. However, inhibition of the histidine and purine metabolism has only been found in microorganisms, unicellular algae and higher plant cells growing in suspension culture but not in intact growing plants [43, 46, 47]. In microorganisms, interference with further biosynthetic pathways, possibly related to those mentioned above, has been found: cysteine biosynthesis was inhibited in *Salmonella typhimuricum* [48], S-adenosylmethionine was found to accumulate in *Eremothecium ashbyii* [49], and coproporphyrin in *Poteriochromonas stipitata* [50]. Also, *aminotriazole* interacts with riboflavin probably by a photoreaction [51,52].

The influence of *aminotriazole* on the soluble amino acid pools in intact higher plants is not well understood. An early effect which has repeatedly been found is a sharp but temporary decrease in the concentrations of free glycine and serine [45]. Other reports have not confirmed this response, but there is more agreement concerning a later increase in the content of all free amino acids [47,53]. The reaction of *aminotriazole* with plant metabolites, which could either proceed by a detoxification mechanism or even involve activation of *aminotriazole* to a phytotoxic metabolite, has extensively been studied [54]. *Aminotriazole* could either react with glucose-1-phosphate to yield a triazolyltriose after aldolase cleavage [55] or be converted to β-(3-amino-1,2,4-triazol-1-yl)-α-alanine [54] via the tryptophan synthetase by substituting for indole [56]. Possibly, one of these metabolites will eventually be incorporated into protein [57]. The possible relationship between the *aminotriazole* metabolism or conjugation reactions and the temporary decrease of the glycine and serine pools is not clear.

Further *aminotriazole* effects include a decreased concentration of IAA which was seen as the reason for lacking geotropic response [58,59]. *Aminotriazole* also inhibits catalase at a concentration of 0.1 M and inactivates this enzyme at 0.02 M in the presence of H_2O_2 [60]. In algae, a suppression of a phosphorylase isoenzyme was detected [61]. On the other hand, the toxicity may be counteracted by reactions with Fe^{3+}, Mg^{2+}, riboflavin or EDTA [34]. The electron transport systems of isolated chloroplasts and mitochondria are not inhibited by 0.1 mM *aminotriazole* [62].

Cell lines resistent to *aminotriazole* have been selected from tobacco callus [63] and suspension culture [64].

b. Fluometuron

Fluometuron is a phenylurea and therefore, as expected, a strong inhibitor of photosynthetic electron transport (p. 31). However, *fluometuron* displays a second mode of action in the development of the functional chloroplast. In *Chlorella pyrenoidosa* [65] and in *Abutilon theophrasti* [66] the light-dependent accumulation of chlorophyll is inhibited, which has been attributed to the inhibition of carotenoid biosynthesis. An increased grana size and a decreased density of stroma lamellae has been observed in *Abutilon*; this may be caused indirectly by the photosynthesis-inhibiting activity of fluometuron.

c. Pyridazinones

Pyridazinone herbicides not only affect the biosynthesis of carotenoids but also other biosynthetic pathways [67]. St. John and Hilton [68] distinguished three different modes of action among the pyridazinones: inhibition of (1) photosynthetic electron transport, (2) carotenoid biosynthesis, and (3) fatty-acid polyunsaturation in the chloroplast galactolipid fraction (Table 28). Fatty-acid dehydrogenations appear to be more generally affected, since they have been reported to be inhibited by pyridazinones e.g. also in roots [69]. The dehydrogenation most commonly inhibited appears to be that of linoleic acid (C 18:2) to linolenic acid (C 18:3):

linoleic acid = C 18:2 =

linolenic acid = C 18:3 =

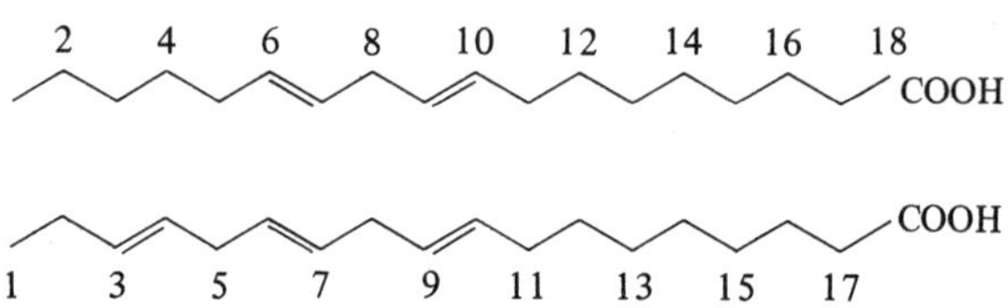

However, the dehydrogenation of C 16:0 to C 16:1 may also be inhibited [70]. Also, in [70] a differential effect of *metflurazone* and SAN 9785 on different lipid fractions has been reported which led the authors to argue that "possibly (different) desaturase enzymes are affected to a different extent by different compounds". Since also in the carotenoid biosynthetic pathway dehydrogenation reactions, occurring in this case in the isoprenoid chain, are affected, inhibition of C-C dehydrogenations appears to be the common mode of action of pyridazinone structures, when photosynthesis inhibition is excluded from the consideration. The structural requirements for each of the three different modes of action may be deduced from Table 28; they have also been dealt with in several publications [70–72].

A very interesting correlation exists between the C 18:2/C 18:3 ratio in the polar lipid fraction and the growth temperature of plants (Table 29). As the growth temperature decreases, the amount of lipid polyunsaturation increases, and this change is inhibited by pyridazinones (68, 69, 73, 74, 74a]. In many cases, the cold-hardiness induced in plants grown at low temperatures was lost together with the increased content of linolenic acid after treatment with pyridazinones [68,

Table 28. Influence of different pyridazinones on photosynthesis (expressed by the pI_{50}-value) [76], on linoleic acid unsaturation (18:2/18:3), and on carotene/chlorophyll synthesis [68]

Pyridazinone structure	pI_{50}	18:2/18:3	Chlorophylls (μg/g dry wt.)	Carotenoids (μg/g dry wt.)
Control	–	0.50	509	42
Pyrazon (Table 11)	5.4	0.57	471	37
SAN 9774 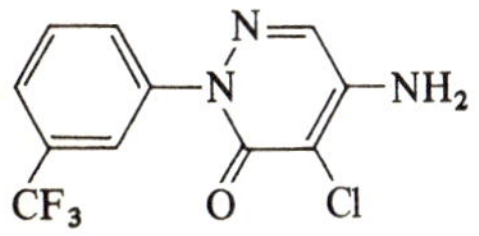	–	0.89	227	24
Norflurazon (Table 26)	4.6	1.83	22	< 1
Metflurazone (Table 26)	5.4	2.47	5	< 1
SAN 9785 =BAS 13–338 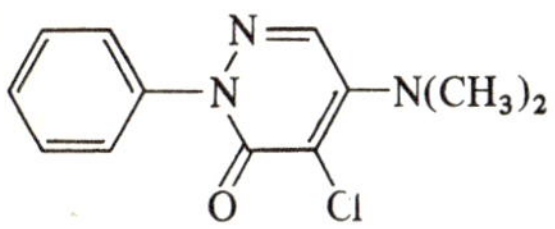	–	4.07	521	41
SAN 113–410H 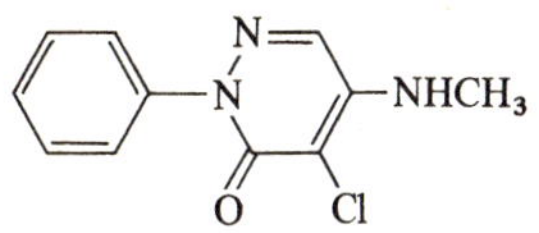	–	1.06	305	25

Table 29. Variation of the linoleic acid/linolenic acid (C 18:2/C 18:3) ratio in cotton seedling roots with the growth temperature and the influence of SAN 9785 [75]

Growth temperature (°C)	Treatment	
	Control	SAN 9785 (10 μM)
30	2.67	4.04
25	1.46	3.08
20	1.18	3.07
15	1.16	2.56

69, 73], but exceptions to this rule have also been observed [74]. Clearly, though linoleic acid unsaturation does appear to play a crucial role in cold-hardiness, it may not be the main factor.

Quinn et al. [77] distinguish chill-hardiness, cold-hardiness, and freeze-resistance, each possibly involving its own mechanism of induction, realization and eventual damage. Chill-hardiness appears to be induced by a water-stress and therefore shows relations to draught hardening and salt resistance [78]. The most important factor in cold-hardiness appears to be the fluidity of the membrane. Damage of the chloroplast upon the action of cold stress occurs below a certain membrane fluidity, and the temperature-dependent adaptation of the membrane fluidity consequently determines the actual temperature range tolerated by the chloroplast [79]. The fluidity of the membrane at a given temperature is higher when the temperature was low during the adaptation phase, thereby shifting the tolerated temperature range to lower degrees. However, though the membrane fluidity generally increases with decreased C 18:2/C 18:3 ratios and increased cold-hardiness [80], lipid polyunsaturation is not always correlated with "frost tolerance" [81].

To sum up, the situation is very complex. The current "fluid mosaic model" of membrane structure envisages an asymmetric distribution of different lipid classes across and in the plane of the membrane [77] (Fig. 10). In this model, the membrane fluidity is of prime importance for the temperature adaptability of the plant membrane. Fatty-acid polyunsaturation appears to contribute only partly to the necessary changes [90]. It has, however, also been found that the membrane fatty-acid desaturase activity is temperature-regulated [91], i.e. induced at low temperatures. Therefore, the hypothesis that the membrane fluidity/fatty acid desaturase-system may be self-regulating has been suggested.

The inhibition of photosynthetic electron transport is one of the three modes of action observed with pyridazinone herbicides [68] (Table 28). However, only *pyrazon* and *metflurazone* appear to have pI_{50}-values possibly causing in vivo inhibition of photosynthesis. Of these two, only *pyrazon* appears to be herbicidally active through the inhibition of photosynthesis (p. 46). Though an inhibition of photosynthesis and $^{14}CO_2$-fixation may be observed after treatment with *metflurazone* [72, 82], this may not be relevant in the field because the compound already hinders the development of the functional chloroplast.

Tolerance against pyridazinone herbicides results from very different mechanisms. *Metflurazone* may simply be trapped in some lipophilic compartments and thereby be restricted in its movement, e.g. by the addition of phytol, farnesol, squalen, unsaturated fatty acids etc. [67]. Such a trapping compartment may be the lysigenous glands of the tolerant cotton plants [83]. In addition, also metabolism appears to be of great importance, especially in sensitive plants where *metflurazone* demethylation occurs at a greater rate [84]. Therefore, the actual phytotoxic species might be the monomethylated *norflurazon* in both cases.

Norflurazon has been used in the study of morphogenesis and phytochrome action in carotenoid free plants [85–89 b]. In general, growth, morphogenesis and phytochrome action were all unaffected. However, PAL (phenylalanine ammonia-lyase) activiy was either increased or decreased in barley and mung bean, depending on concentration and organ [85, 86]. Moreover, the levels of the 3-hydroxylated flavonoids saponarin and lutonarin were increased in barley seedlings. The interpretation given by these authors suggests that "the absence of blue light absorbing carotenoids in herbicide treated plants allows more complete expression of the blue light High Energy photoresponse". Also, nitrate reduction has been studied in *norflurazon* treated tissue [89 a].

d. Haloxydine and Pyriclor

These two herbicides exhibit a mode of action differing from that of the "inhibitors of carotenoid biosynthesis". *Haloxydine* [20, 21] and *pyriclor* [93] both act on established green tissue in the light. In this respect, they may be compared with diphenyl ether herbicides. With these herbicides they also share the action on chlorophyll-free but carotenoid – containing tissue [20]. In contrast to the "true inhibitors of carotenoid biosynthesis", *haloxydine* has been shown to inhibit the greening of etiolated tissue. However, when plants are grown in the light in the presence of *haloxydine* or *pyriclor*, the tissue is devoid of carotenoids but accumulates phytoene, phytofluene, and ζ-carotene [3, 21]. An interference between carotenoids and these herbicides in the light might be inferred from these results, which, however, appears to be different from the primary mode of action of both the "inhibitors of carotenoid biosynthesis" and diphenyl ether herbicides.

Data on the action of *pyriclor* on the electron transport and phosphorylation systems of chloroplasts [95] and mitochondria have been reported [94]. Inhibitions only occurred above 10 μM *pyrichlor*. Gruenhagen et al. [96] have reported on the ATP-level in etiolated soybean hypocotyls which was decreased in the presence of 100 μM *pyriclor*.

e. Others

Detailed information on the action of herbicides on physiological and biochemical reactions of plants is usually only available for a few selected structures. Also, in this group many compounds, which e.g. include *ioxysone* [97] and *U 18* [98], have only broadly been described as "bleaching herbicides". They might, of course, display a different primary mode of action.

Difunon has been analyzed for its possible influence on the succinyl – CoA – and δ-aminolevulinic acid synthetases, but none has been found [99]. Photosynthesis and respiration have been reported to be insensitive to *methoxyphenone* [100, 101] but to be inhibited in isolated mesophyll cells by *fluridone* [102] at a concentration of 0.1 mM. The tolerance of plants for *methoxyphenone* has been studied by Fujii et al. [15, 101]. Since they could not find differences in the uptake, transport and detoxification rates of *methoxyphenone*, the authors discuss possible differences of the sensitive enzyme systems among tolerant and susceptible plants.

References for C. XI

1. Wolf, F.T.: The Bot. Rev. *43*, 395 (1977)
2. Goodwin, T.W.: Carotenoids. In: Phytochemistry, Vol. I. Miller, L.P. (ed.), p. 112. New York, Cincinnati, Toronto, London, Melbourne: Van Nostrand Reinhold Comp. 1973
3. Burns, E.R. et al.: Plant Physiol. *47*, 144 (1971)
4. Hilton, J.L.: J. Agr. Food Chem. *17*, 182 (1969)
5. Rüdiger, W., Benz, J.: Z. Naturforsch. *34c*, 1055 (1979)
6. Urbach, D. et al.: Z. Naturforsch. *31c*, 652 (1976)
7. Kunert, K.J., Böger, P.: Weed Sci. *26*, 292 (1978)

8. Bartels, P.G., Watson, C.N.: Weed Sci. *26*, 198 (1978)
9. Bartels, P.G., McCullough, C.: Biochem. Biophys. Res. Comm. *48*, 16 (1972)
10. Ben-Aziz, A., Koren, E.: Plant Physiol. *54*, 916 (1974)
11. Kümmel, H.W., Grimme, L.H.: Z. Naturforsch. *30c*, 333 (1975)
12. Lichtenthaler, K.H., Kleudgen, H.K.: Z. Naturforsch. *32c*, 236 (1977)
13. Rüdiger, W. et al.: Z. Pflanzenphysiol. *80*, 131 (1976)
14. Vaisberg, A.J., Schiff, J.A.: Plant Physiol. *57*, 260 (1976)
15. Fujii, Y. et al.: J. Pestic. Sci. *3*, 291 (1978)
16. Ridley, S.M., Ridley, J.: Plant Physiol. *63*, 392 (1979)
17. Fujii, Y. et al.: J. Pestic. Sci. *2*, 431 (1977)
17a. Sandmann, G., Bramley, P.M., Böger, P.: Pestic. Biochem. Physiol. *14*, 185 (1980)
18. Kleinig, H.: Arch. Microbiol. *97*, 217 (1974)
18a. Beyer, B., Kreuz, K., Kleinig, H.: Planta *150*, 435 (1980)
19. Murillo, F.J.: Plant Sci. Lett. *17*, 201 (1980)
20. Dodge, J.D., Lawes, G.B.: Ann. Bot. *36*, 315 (1972)
21. Feierabend, J. et al.: Z. Naturforsch. *34c*, 1036 (1979)
22. Devlin, R.M. et al.: Weed Sci. *26*, 432 (1978)
23. Bartels, P.G., Hyde, A.: Plant Physiol. *45*, 807 (1970)
24. Devlin, R.M., Kisiel, M.J., Karczmarczyk, S.J.: Weed Res. *16*, 125 (1976)
25. Bartels, P.G. et al.: Plant Physiol. *42*, 736 (1967)
26. Sankhla, N., Ziegler, H.: Naturwiss. *60*, 157 (1973)
27. Favali, M.A., Conti, G.G.: Protoplasma *70*, 153 (1970)
28. Krinsky, N.J.: Carotenoid pigments: multiple mechanisms for coping with the stress of photosensitized oxidations. In: Strategies of microbial life in extreme environments. Shilo, M. (ed.), p. 163. Weinheim, New York: Verlag Chemie 1978
29. Searle, G.F.W., Wessels, J.S.C.: Biochim. Biophys. Acta *504*, 84 (1978)
30. Takahama, U.: Plant Cell Physiol. *19*, 1565 (1978)
31. Halliwell, B.: Prog. Biophys. Molec. Biol. *33*, 1 (1978)
32. Kunert, K.J., Böger, P.: Z. Naturforsch. *34c*, 1047 (1979)
33. Kulandaivelu, G., Gnanam, A.: Physiol. Plant. *33*, 234 (1975)
34. Vivekandanan, M., Gnanam, A.: Plant Physiol. *55*, 526 (1975)
35. Ito, K. et al.: Some aspects of mode of action of new herbicide NK-049. In: Proc. fifth Asian Pacific Weed Sci. Soc. Conf., p. 159 (1975)
36. Geronimo, J., Herr, J.W.: Weed Sci. *18*, 48 (1970)
37. Appleman D.: Plant Physiol. *27*, 613 (1952)
38. Feierabend, J., Beevers, H.: Plant Physiol. *49*, 28 (1972)
39. Feierabend, J., Beevers, H.: Plant Physiol. *49*, 33 (1972)
40. Feierabend, J.: Planta *123*, 63 (1975)
41. Feierabend, J., Schubert, B.: Plant Physiol. *61*, 1017 (1978)
42. Blume, D.E., McClure, J.W.: Plant Physiol. *65*, 238 (1980)
43. Bartels, P., Wolf, F.T.: Physiol. Plant. *18*, 805 (1965)
44. Davies, M.E.: Phytochem. *10*, 783 (1971)
45. Burt, G.W., Muzik, T.J.: Physiol. Plant. *23*, 498 (1970)
46. Baumann, G., Günther, G.: Biochem. Physiol. Pflanzen *167*, 399 (1975)
47. Baumann, G., Günther, G.: Biochem. Physiol. Pflanzen *169*, 163 (1976)
48. Hulanicka, D.: J. Gen. Microbiol. *72*, 291 (1972)
49. Suzuki, Y., Emoto, H., Mitsuda, H.: Biochim. Biophys. Acta *541*, 115 (1978)
50. Dörfling, P., Dummler, W., Mucke, D.: Experientia *26*, 728 (1970)
51. Naylor, A.W., Hilton, J.C.: J. Agr. Food Chem. *12*, 21 (1964)
52. Baumann, G., Günther, G.: Biochem. Physiol. Pflanzen *169*, 171 (1976)
53. Suen, R.S.K., Lenoir, J.A., Ibrahim, R.K.: Z. Pflanzenphysiol. *93*, 31 (1979)
54. Lund-Hoie, K.: Weed Res. *10*, 67 (1970)
55. Fredrick, J.F., Gentile, H.C.: Ann. of the New York Acad. Sci. *144*, 362 (1967)
56. Smith, L.W., Chang, F.Y.: Weed Res. *13*, 339 (1973)
57. Brown, J.C., Carter, M.C.: Weed Sci. *16*, 222 (1968)
58. Burt, G.W., Muzik, T.J.: Plant Physiol. *47*, 795 (1971)
59. Counts, B.: Weeds 9, 329 (1961)

60. Margolish, E., Novogrodsky, A., Schejter, A.: Biochem. J. *74*, 339 (1960)
61. Fredrick, J.F.: Phytochem. *8*, 725 (1969)
62. Foy, C.L., Penner, D.: Weeds *13*, 226 (1965)
63. Barg, R., Umiel, N.: Z. Pflanzenphysiol. *83*, 437 (1977)
64. Richter, P.L., McDaniel, C.N.: Plant Physiol. Suppl. *63*, no. 819 (1979)
65. Sikka, H.C., Pramer, D.: Weed Sci. *16*, 296 (1968)
66. Wergin, W.P., Potter, J.R.: Pestic. Biochem. Physiol. *5*, 265 (1975)
67. Hilton, J.L., St. John, J.B., Christiansen, M.N.: Plant Physiol. *48*, 171 (1971)
68. St. John, J.B., Hilton, J.L.: Weed Sci. *24*, 579 (1976)
69. Willemot, C.: Plant Physiol. *60*, 1 (1977)
70. Khan, M.-U. et al.: Plant Physiol. *64*, 300 (1979)
71. Urbach, D. et al.: Z. Naturforsch. *31c*, 652 (1976)
72. Hilton, J.L. et al.: Weed Sci. *17*, 541 (1969)
73. St. John, J.B.: Chemical modification of lipid in chilling sensitive plants. In: Low temperature stress in crop plants – the role of the membrane. Lyons, J.M., Graham, D., Raison, J.K. (eds.), p. 405. London, New York, San Francisco: Acad. Press 1979
74. de la Roche, A.J.: Plant Physiol. *63*, 5 (1979)
74.a. Ashworth, E.N. et al.: Plant Physiol. *67*, 711 (1981)
75. St. John, J.B. Christiansen, M.N.: Plant Physiol. *57*, 257 (1976)
76. Trebst, A., Harth, E.: Z. Naturforsch. *29c*, 232 (1974)
77. Quinn, P.J., Williams, W.P.: Prog. Biophys. Molec. Biol. *34*, 109 (1978)
78. Stuiver, C.E.E., Kuiper, P.J.C., Marschner, H.: Physiol. Plant. *42*, 124 (1978)
79. Raison, J., Berry, J., Björkman, O.: Plant Physiol. Suppl. *63*, no. 779 (1979)
80. Wang, C.Y., Baker, J.E.: Plant Cell Physiol. *20*, 243 (1979)
81. Vigh, L. et al.: Phytochem. *18*, 787 (1979)
82. Porter, E.M., Bartels, P.G.: Weed Sci. *25*, 60 (1977)
83. Strang, R.H., Rogers, R.L.: Weed Sci. *23*, 26 (1975)
84. Strang, R.H., Rogers, R.L.: Weed Sci. *22*, 1119 (1974)
85. Blume, D.E., McClure, J.W.: Phytochem. *17*, 1545 (1978)
86. Blume, D.E., McClure, J.W.: Phytochem. *17*, 1549 (1978)
87. Jabben, M., Deitzer, G.F.: Plant Physiol. *63*, 481 (1979)
88. Frosch, S. et al.: Planta *145*, 497 (1979)
89. Gorton, H.L., Briggs, W.R.: Plant Physiol. *66*, 1024 (1980)
89a. Widell, K.-O., Sundqvist, C.: Physiol. Plant. *52*, 325 (1981)
89b. Grill, R., Schraudolf, H.: Plant Physiol. *68*, 1 (1981)
90. Markhart, A.H. et al.: Plant Cell Env. *3*, 435 (1980)
91. Thompson, Jr., G.A.: Molecular control of membrane fluidity. In: Low temperature stress in crop plants – The role of the membrane. Lyons, J.M., Graham, D., Raison, J.K. (eds.), p. 347. London, New York, San Francisco: Academic Press 1979
92. Öquist, G., Samuelsson, G., Bishop, N.J.: Physiol. Plant. *50*, 63 (1980)
93. Geronimo, J., Herr, J.W.: Weed Sci. *18*, 48 (1970)
94. Killion, D.D., Frans, R.E.: Weed Sci. *17*, 468 (1969)
95. Meikle, R.W.: Weed Sci. *18*, 475 (1970)
96. Gruenhagen, R.D. Moreland, D.E.: Weeds *19*, 319 (1971)
97. Sarma, R., Flood, A.E., Wain, R.L.: Ann. Appl. Biol. *86*, 429 (1977)
98. Takamura, S., Matsunaka, S.: Weed Abstr. *22*, 284 (1973)
99. Hampp, R., Sankhla, N., Huber, W.: Physiol. Plant. *33*, 53 (1975)
100. Fujii, Y. et al.: J. Pestic. Sci. *4*, 391 (1979)
101. Fujii, Y. et al.: J. Pestic. Sci. *4*, 511 (1979)
102. Rafii, Z.E.: Weed Sci. *27*, 422 (1979)

D. Energy Conservation

I. Energy Conservation in Photosynthesis and Respiration

Historically, three different types of biological energy conservation systems may be distinguished: oxidative phosphorylation, photosynthetic phosphorylation, and substrate phosphorylation. All these systems conserve part of the free energy obtained from photochemical or biochemical processes in the form of ATP (adenosine triphosphate). For the present consideration only the oxidative and photosynthetic phosphorylation systems need to be discussed.

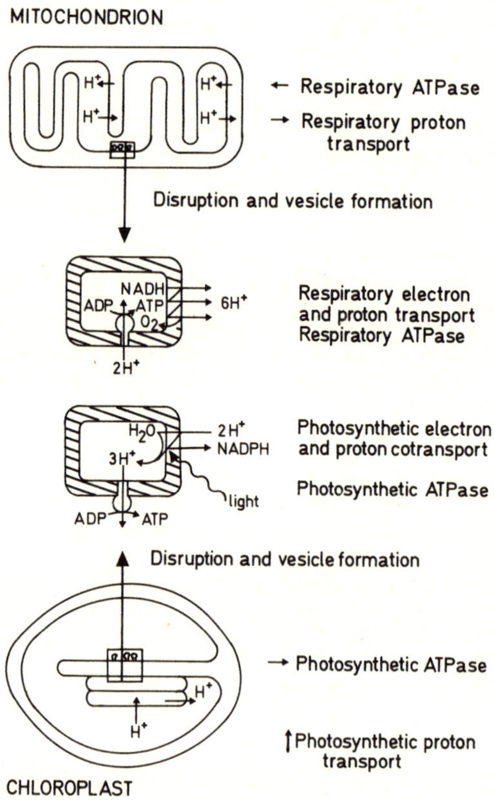

Fig. 36. Vectorial electron flow and energy conservation in mitochondria and chloroplasts. The arrows indicate the direction of proton transport in intact organelles. Details not drawn to scale. Adapted from Hinkle et al. [1]

The photosynthetic and the oxidative phosphorylation systems can be understood by the chemiosmotic theory (Fig. 36). Basically, a vectorial (directed) electron transport system pumps protons through a membrane, thus rendering the interior acidic [ΔpH between 1.4 (mitochondria) and 3.5 (chloroplast)]. This interior is the space between the outer and inner membrane in mitochondria and the thylacoid interior in chloroplasts. After organelle disruption and suborganelle vesicle formation, mitocondrial particles become alkaline in the interior whereas chloroplast particles become acidic in the interior upon substrate- or light-induced electron transport, respectively; i.e. the chloroplast particles are "inside-out" with respect to the mitochondrial particles, and vice versa.

The synthesis of ATP is catalyzed by an ATPase which contains several coupling factors; these may be seen in electron micrographs as "knobs" on the membrane. During the function of the ATPase, protons are pumped back through the membrane following the prevalent proton gradient downhill, i.e. the pH-gradient supplies the energy needed for the synthesis of ATP. Due to this principle, electron transport and ATP-formation in chloroplasts and mitochondria are not molecularly coupled, but are situated in the membrane at different locations.

II. Uncouplers of Oxidative Phosphorylation

Uncoupling means that, while the electron transport rate remains more or less unchanged, the synthesis of ATP is severely inhibited. The proposed mechanism for an uncoupling of electron transport and phosphorylation by herbicidal uncouplers may be easily understood in terms of the chemiosmotic theory. The herbicides classified here as "respiratory uncouplers" (benzonitriles see Table 12, dinitrophenols see Table 13, benzimidazoles see Table 14, and *pentachlorophenol*) contain a labile phenolic hydrogen or an acidic NH-group in the molecule. They may therefore transport protons through membranes and degrade an existing pH-gradient. The result would be a sharp drop in the rate of ATP-formation with no effect on electron transport, i.e. an uncoupling activity. The requirement of an acidic NH-group to induce the uncoupling effect has also been demonstrated for a series of 2-anilino-1,3,4-thiadiazoles [2]. An optimum pk_a-value of 6.8 was found for the uncoupling of oxidative phosphorylation for this series of compounds. Basically, the same has been observed for the uncoupling of oxidative phosphorylation by 2-anilinothiophenes [3] and for the uncoupling of photosynthetic phosphorylation by 2-alkylbenzimidazoles [4].

The relationship between uncoupling and inhibition of mitochondrial electron transport deserves special attention. With isolated mitochondria, uncoupling may preferably be measured in "state 4", whereas inhibition should be measured in "state 3" (Table 30). "State 4" is called the "control state" where lack of ADP limits the respiratory electron transport. Any uncoupling compound will therefore increase the rate of oxygen uptake in "state 4" mitochondria. This may also occur in intact plants [5]. However, none of the compounds mentioned above is a "pure uncoupler". Instead, they uncouple at lower and inhibit electron transport at somewhat higher concentrations. Moreland has therefore termed these compounds "inhibitory uncouplers" [6, 7]. Inhibitory concentrations needed for respiratory

Table 30. Conditions used for the testing of mitochondrial preparations

Nomenclature	Test deficiency	Limiting factor	Respiration rate
State 1	ADP + Substrate	ADP	Low
State 2	Substrate	Substrate	Low
State 3	None	Electron transport chain	High
State 4	ADP	ADP	Low
State 5	O_2	O_2	Low

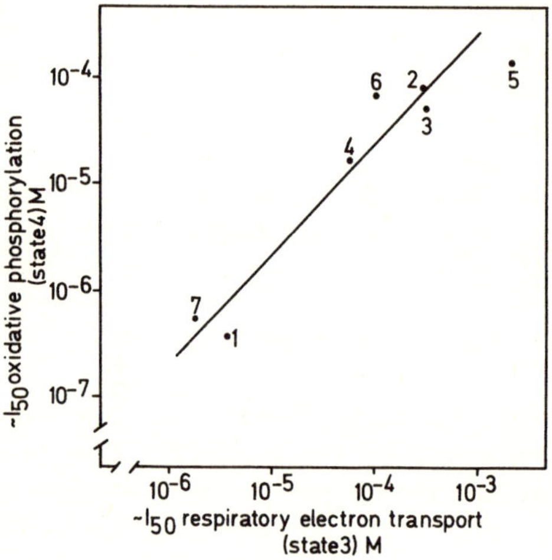

Fig. 37. Comparison of the inhibition of mitochondrial electron transport with the uncoupling of oxidative phosphorylation for some phenolic and NH-acidic compounds. *1 ioxynil, 2 bromoxynil, 3 PCP, 4* 4-OH-dichlobenil, *5* 2,4-dinitrophenol, *6* 3-OH-dichlobenil, *7 dinoseb*

electron transport or oxidative phosphorylation are compared in Fig. 37. Where I_{50} values were not reported, equivalent data were used. Figure 37 illustrates that in mitochondria phosphorylation is between 2 and 10 times more sensitive to many inhibitors (uncouplers) than electron transport. This situation is in sharp contrast to the sensitivity of chloroplasts, which show the opposite characteristics with respect to normal photosystem II inhibitors (Fig. 18). However, 2-alkylbenzimidazoles uncouple photosynthetic phosphorylation without inhibiting photosynthetic electron transport [4]. Conversely, the respiratory uncouplers are also fairly good inhibitors of the photosynthetic electron transport system (Tables 12–14). This inhibitory activity has been discussed in detail (pp. 53, 58).

The higher sensitivity of oxidative phosphorylation towards uncoupling activities of herbicides as compared to the lower sensitivity of the respiratory electron transport chain to herbicides might be a more general phenomenon. This seems to be true not only for the uncouplers mentioned before (Fig. 37) but also for a large number of other herbicides differing widely in their structure as well as in their known primary mode of action (e.g. *fluometuron*, *ametryne*, *prometryne*, diphenyl ether herbicides, *CIPC, trifluralin, 2,4,5-T*). The mode of the uncoupling activity is not clear in these cases. Equally open to discussion is the contribution of a possible in vivo-uncoupling of respiratory energy conservation to the herbicidal activity in the case of these "other herbicides". One approach to the understanding of "unspecific" uncoupler effects has been pointed out by Moreland: membrane perturbations as a result of herbicidal interactions may change the fluidity and other characteristics of the membranes, eventually resulting in partial uncoupling [8]. Depending on time, type and concentration of the application, many herbicides may therefore display some uncoupling activity in intact tissue.

III. Respiratory Uncoupling in the Intact Tissue

It should be stressed that in intact plant tissue two different types of respiration may occur: a "normal respiration" and an "alternate respiration" [32]. Whereas the "normal respiration" shows only a comparatively (as compared with animal mitochondria) poor coupling of ADP-phosphorylation with the respiratory electron transport, the "alternate respiration" – via an "alternate oxidase" – ist not coupled at all. The uncoupling activity may therefore, depending on the type of tissue and respiration, vary considerably with experiments. In general, seedling or other proliferating tissue with high respiratory rates will respond best. *Ioxynil* and, to a lesser extent, *neburon*, *chlorfenprop-methyl*, and *DNOC* have been found to inhibit the alternate pathway [9].

In etiolated tissue uncouplers induce a severe drop in the cellular ATP supply [10]. As a consequence, metabolic processes directly depending on a continued ATP supply suffer first. Among these, protein synthesis [11], glyceride synthesis [12], and mitosis [5] are some of the more prominent. The damage occurring in the tissue will also increase the permeability of the membranes [13].

In the intact green plant the situation may be more complex. Respiratory uncoupling may be one of the early effects, but inhibition of respiration, photosynthesis, and transpiration soon become prominent [14, 15]. The inhibition of transpiration might limit further uptake of the herbicide, thereby limiting the extent of e.g. photosynthesis inhibition [16]. The relative contributions of photosynthetic and respiratory inhibitions by the various herbicides may differ. The results obtained by van Oorschot [15] suggest that for *DNOC* the inhibition of respiration may be more important whereas for *ioxynil* the opposite may be true. However, this might be different in other systems [16, 17].

To sum up, "respiratory uncouplers" exhibit at least two primary modes of action in photosynthesis and respiration. They also rapidly and severely affect indirectly every stage of plant metabolism due to a "cellular energy crisis". The relative importance of the two primary modes of action for the herbicidal activity depends on the specific tissue and its physiological situation.

3,5-Diiodo-4-hydroxybenzoic acid (DIHB), the free acid corresponding to *ioxynil*, has been found to inhibit auxin-induced ethylene formation [18, 19]. DIHB is formed from *ioxynil* in the soil and may then enter the plant. DIHB may be compared with other halogenated benzoic acids showing auxin activity (p. 161) or auxin-inhibitor activity (p. 177).

IV. Dichlobenil – an Inhibitor of Cellulose Biosynthesis

The herbicide *dichlobenil* is a poor uncoupler of oxidative and photosynthetic phosphorylation systems, and an even poorer inhibitor of respiratory and photosynthetic electron transport [20, 21]. However, the hydroxylation products formed in the plant have been found to be fairly potent uncouplers (Table 31) [21, 22]. It has therefore been assumed that *dichlobenil* may be herbicidally active via its metabolites. However, Gruenhagen and Moreland [10] did not find changes in the ATP concentration of etiolated soybean tissue treated with *dichlobenil*. In order to answer the question of the possible importance of the hydroxylated *dichlobenil* metabolites for the herbicidal activity, physiological experiments should consequently be combined with a monitoring of the herbicidal metabolism.

The herbicide *chlorthiamid* is converted to *dichlobenil* in the soil [23, 24] or under the influence of light [25].

The presumptive primary mode of action of *dichlobenil* has been detected in an interesting series of investigations. Firstly, experiments with ^{14}C-glucose have shown an inhibition of cellulose labelling in "Azuki" bean epicotyls [26]. This inhibition was not accompanied by a decrease of the cortical microtubules (cf. p. 128) and therefore presumably represented a more direct metabolic effect. Umetsu et al. [27] observed irregularly swollen cells in soybean cell cultures which they compared with cells treated with coumarin, an inhibitor of cellulose synthesis. Meyer et al. [28] observed that in tobacco callus cells the formation of new cell walls was completely suppressed by 1 $mg\,l^{-1}$ *dichlobenil*. However, nuclear divisions and

Table 31. *Chlorthiamid, dichlobenil,* and some hydroxylation products. The data give the I_{50}-values for the uncoupling of open-chain photophosphorylation (P) and optimum concentrations for the uncoupling of "state 4"-respiration (R) in mitochondria [21] in molar

Common name	Structure	Uncoupling of	
		P	R
Chlorthiamid	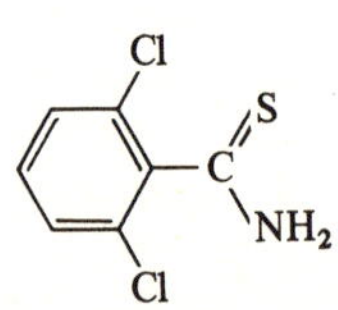	–	–
Dichlobenil	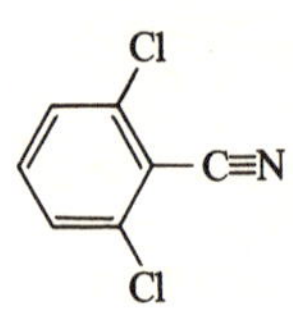	$>10^{-3}$	10^{-3}
3-OH-dichlobenil	–	6.2×10^{-5}	5.0×10^{-5}
4-OH-dichlobenil	–	1.8×10^{-5}	1.5×10^{-5}

cytokinesis were not inhibited but only secondarily affected, yielding polynucleate cells. Montezinos et al. [29] finally showed that the formation of β 1→4 glucane was selectively inhibited in cotton ovules by this herbicide. The formation of β 1→3 glucane was only 50% sensitive, and the formation of other polysaccharides was insensitive to 10 μM *dichlobenil.*

Another observation which has been made in *dichlobenil*-treated tissue is an increase of the activity of the enzyme phenylalanine ammonia lyase (PAL) (cf. p. 185) [30, 31]. Simultaneously, the cellular cinnamic acid pools were depleted. Since PAL is an inducible enzyme, the increase possibly results from a feed-back control. It is not known whether the effect on PAL activity is a consequence of the inhibited cell wall formation.

The water solubility of *dichlobenil* is 18 $mg l^{-1}$.

References for D

1. Hinkle, P.C., McCarty, R.E.: Scientific American *238*, 104 (1978)
2. Schäfer, G., Trebst, A., Büchel, K.H.: Z. Naturforsch. *30c*, 183 (1975)
3. Büchel, K.H., Schäfer, G.: Z. Naturforsch. *25b*, 1465 (1970)
4. Trebst, A., Pistorius, E., Büchel, K.H.: Z. Naturforsch. *21b*, 667 (1966)
5. Wakamori, S.: Agr. Biol. Chem. *37*, 307 (1973)
6. Alsop, W.R., Moreland, D.E.: Pestic. Biochem. Physiol. *5*, 163 (1975)
7. Moreland, D.E.: Mode of action of herbicides. In: Pesticide chemistry in the 20th century. Plimmer, J. (ed.), p. 56. American Chemical Soc. 1977
8. Moreland, D.E.: Ann. Rev. Plant Physiol. *31*, 597 (1980)
9. Blein, J.P.: Plant Sci. Lett. *19*, 65 (1980)
10. Gruenhagen, R.D., Moreland, D.E.: Weeds *19*, 319 (1971)
11. Mann, J.D., Jordan, L.S., Day, B.E.: Plant Physiol. *40*, 840 (1965)
12. St. John, J.B., Hilton, J.L.: Weed Sci. *21*, 477 (1973)
13. Prendeville, G.N., Warren, G.F.: Weed Res. *30*, 251 (1977)
14. van Oorschot, J.L.P.: Pestic. Sci. *1*, 33 (1970)
15. van Oorschot, J.L.P.: Acta Bot. Neerl. *23*, 36 (1974)
16. Moreland, D.E., Hilton, J.L.: Actions on photosynthetic systems. In: Herbicides – physiology, biochemistry, ecology, Vol. 1. Audus, J.L. (ed.), p. 493. London, New York, San Francisco: Academic Press 1976
17. Berüter, J., Temperli, A.T.: Experientia *26*, 600 (1970)
18. Robert, M.L., Taylor, H.F., Wain, R.L.: Planta *129*, 53 (1976)
19. Larque-Saavedra, A., Wilkins, H., Wain, R.L.: Planta *126*, 269 (1975)
20. Foy, C.L., Penner, D.: Weeds *13*, 226 (1965)
21. Moreland, D.E., Hussey, G.G., Farmer, F.S.: Pestic. Biochem. Physiol. *4*, 356 (1974)
22. Frear, D.S. et al.: Advances in Agron. *24*, 327 (1972)
23. Milborrow, B.V.: Biochem. J. *87*, 255 (1963)
24. Milborrow, B.V.: Weed Res. *5*, 332 (1965)
25. Bahadir, M. et al.: J. Agr. Food Chem. *27*, 815 (1979)
26. Hogetsu, T., Shibaoka, H., Shimokoriyama, M.: Plant Cell Physiol. *15*, 389 (1974)
27. Umetsu, N., Satoh, S., Matsuda, K.: Plant Cell Physiol. *17*, 1071 (1976)
28. Meyer, Y., Herth, W.: Planta *142*, 253 (1978)
29. Montezinos, D., Delmer, D.P.: Planta *148*, 305 (1980)
30. Engelsma, G.: Acta Bot. Neerl. *22*, 49 (1973)
31. Amrhein, N., Gerhardt, J.: Biochim. Biophys. Acta *583*, 434 (1979)
32. Lambers, H.: Plant Cell Envir. *3*, 293 (1980)

E. Nucleic Acid and Protein Synthesis

The syntheses of nucleic acids and proteins in plants have, with two exceptions, not been proven to be primary targets for the action of herbicides. However, since the syntheses of nucleic acids (RNA, DNA) and proteins depend on a continuous supply of metabolic energy in the form of ATP, they might be inhibited secondarily. In effect, when studying the available data on the influence of herbicides on RNA and protein synthesis in the dark (Table 32), the most inhibitory compounds are found to be *dinoseb*, *ioxynil*, and *pyrichlor*, and these are well-known uncouplers of oxidative phosphorylation. Other compounds like e.g. *CDAA*, *chlorpropham*, *endothal*, *propachlor*, and *2,4,5-T* may inhibit the syntheses of these polymers as a consequence of another primary metabolic interference (e.g. for *CDAA* and *propachlor* [4]). However, at the high concentrations employed in some of these experiments, partial uncoupling may also be induced by many herbicides (p. 116) which exhibit a different mode of action at lower concentrations in other systems (e.g. *CIPC*, *diuron*, *EPTC*, *propachlor*, *propanil*, *2,4,5-T*, *trifluralin*). Similarly, photosynthesis-inhibiting herbicides also indirectly inhibit other metabolic processes, in this case in the light (e.g. *atrazine*, *bromacil*, *monuron*). To sum up, the inhibition by herbicides of in vivo systems synthesizing proteins or nucleic acids may usually be regarded as a secondary process.

One of the exceptions mentioned above is *MDMP* (Table 33), a herbicide which has not been marketed but which has been shown to exert a direct effect on protein synthesis in higher plants [5, 6]. The precise mode of action of *MDMP* has been reported to be a specific block of the interaction of 60 S ribosomal subunits with the 40 S ribosomal subunit -RNA-Met-tRNA complex [6]. The consequence is an inhibition of the initiation of peptide-chain synthesis; peptide-chain elongation is not affected. *MDMP* has subsequently been used to answer specific physiological questions. By using *MDMP*, it has e.g. been confirmed that protein synthesis is essential for the initial action of auxins in the stimulation of growth [7]. Furthermore, questions regarding the subcellular location of the synthesis of certain proteins can be investigated by use of *MDMP*, since this compound only interferes with cytoplasmic protein synthesis [8]. A retardation of senescence in the presence of *MDMP* has been observed by Thomas [9].

The second possible exception is the new herbicide *DPX 4189*, which has been shown to specifically inhibit DNA-synthesis in corn roots. A more detailed description of this herbicide is given elswhere (p. 156).

Table 32. Influence of herbicides on the syntheses of proteins and nucleic acids in intact plant tissue (negative values = stimulation).
Dark: conc. a = average of 2 and 5 mg l^{-1} (1)
conc. b = 6×10^{-4} M (2)
conc. c = 2×10^{-4} M (2)
Light: conc. d = average of 10^{-5} M and 10^{-4} M (3)

Herbicide	Concentration	Inhibition (%)			
		Protein synthesis		RNA synthesis	
		Dark	Light	Dark	Light
Amiben	a, d	2	31	–	46
	b	3	–	−19	–
Amitrol	a	9	–	–	–
	b	− 3	–	−25	–
Atrazine	a, d	22	19	–	70
	c	14	–	− 6	–
Bromacil	d	–	16	–	71
CDAA	a, d	67	50	–	46
	b	12	–	−17	–
CDEC	a	32	–	–	–
	c	33	–	35	–
Chlorpropham	a, d	60	19	–	27
	b	89	–	77	–
2,4-D	a, d	10	27	–	42
Dacthal	a	22	–	–	–
Dalapon	a, d	7	10	–	42
Dichlobenil	a, d	32	16	–	38
	c	33	–	23	–
Dinoseb	b	98	–	86	–
	d	–	98	–	97
Diphenamid	a	12	–	–	–
Diuron	b	42	–	50	–
Endothal	a	45	–	–	–
EPTC	a, d	21	57	–	63
	b	24	–	22	–
Ioxynil	a	71	–	–	–
	c	97	–	74	–
Monuron	a, d	25	49	–	79
Paraquat	d	–	24	–	71
PCP	a	46	–	–	–
Picloram	b	− 1	–	− 8	–
Propachlor	b	44	–	28	–
Propanil	a	5	–	–	–
	b	90	–	66	–
Pyrazon	a	29	–	–	–
Pyrichlor	b	88	–	68	–
2,4,5-T	b	67	–	44	–
Trifluralin	a, d	13	58	–	70
	c	21	–	3	–

Table 33. Structure of the herbicidal protein synthesis inhibitor *MDMP*

MDMP

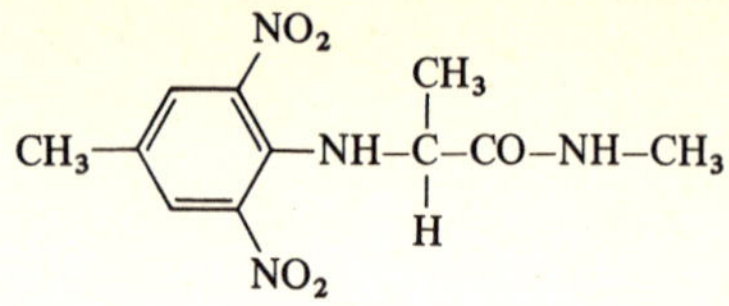

References for E

1. Mann, J.D., Jordan, L.S., Day, B.E.: Plant Physiol. *40*, 840 (1965)
2. Moreland, D.E. et al.: Weed Sci. *17*, 556 (1969)
3. Ashton, F.M. et al.: Pestic. Biochem. Physiol. *7*, 122 (1977)
4. Deal, L.M. et al.: Weed Sci. *28*, 334 (1980)
5. Kerr, W.M., Averey, R.: Biochem. J. Proc. *128*, 132 (1972)
6. Weeks, D.P., Baxter, R.: Biochem. *11*, 3060 (1972)
7. Bates, G.W., Cleland, R.E.: Planta *145*, 437 (1979)
8. Ellis, R.J.: Phytochem. *14*, 89 (1975)
9. Thomas, H.: Plant Sci. Lett. *6*, 369 (1976)

F. Microtubules

I. Microtubules in the Cell Cycle of Higher Plants

The mystery of plant morphogenesis has occupied biologists and specifically botanists since generations. In recent years our understanding of plant morphogenesis has been considerably advanced through the analysis of the organization and function of microtubules in plant cells.

Microtubules appear in electron micrographs as elongated narrow structures or, when cross-sectioned, as hollow rings. They are hollow and flexible cylinders up to 5.2 μm long and 25 nm in diameter. They occur at different locations and at different times in the cell during its life cycle [5–7] and are composed of the protein tubulin which itself is a dimer.

A scheme describing the microtubular functions during the life cycle of a higher plant cell is given in Table 34. This scheme is based on data from Gunning et al. [5] which were mostly obtained from studies with the fern *Azolla*; however, according to the present knowledge the principles outlined should also be true for a generalized higher plant cell. The basic principles for understanding the scheme are (a) reversible polymerization/depolymerization of tubulin into microtubules (MTs) at definite times during the cellular life cycle and at specific locations in the cell and (b) control of the polymerization/depolymerization process by microtubular organizing centers (MTOCs) in these different locations. This means that a pool of free tubulin persists in the cytoplasm which is reversibly polymerized to microtubules by the appropriate MTOCs.

When starting the life of a new cell with the nuclear division in the mother cell, the microtubules to be considered first are the "spindle microtubules". Higher plants do not possess a centriole, the distinct structure from which the spindle microtubules originate in lower plants and animals. Instead, in higher plants the spindle microtubules originate from an electron dense cytoplasmic area. Some spindle microtubules are attached to the kinetochores of the individual chromosomes, and chromosome displacement to the spindle poles subsequently takes place (Anaphase). Likewise, the other MTOC's outlined in Table 34 also appear as electron dense areas of the cytoplasm when viewed in an electron microscope.

After the daughter nuclei have formed and a new primary cell wall has been synthesized at the site of the phragmoplast, a secondary cell wall is produced by the apposition of cellulose microfibrils. This often occurs in ordered arrays of different layers with crossed textures. Interestingly, the direction of growing cellulose microfibrils is parallel to microtubules located subjacent to the plasmalemma and shows bridges to the latter. These so-called "cortical microtubules" are thought to

Table 34. Microtubular functions during the cell cycle of a higher plant, based on results from Gunning et al. [5]. MT = microtubule, T = tubulin, MTOC = microtubule-organizing center

	Metaphase	Anaphase		Telophase		Interphase		Prophase
→ T	——————→	Spindle MTs	——→ T ——→	cell plate MTs	——→ T ——→	cortical MTs	——→ T ——→	pre-prophase band MTs → T →
		⇅		⇅		⇅		⇅
MTOC		Spindlepole, Centriole		Phragmoplast		Matrix at cell edges		Formation of new matrix
Function		Chromosome distribution in nuclear divisions		Formation of new primary cell wall		Cell-wall differentiation, orientation, and patterning of cellulose microfibrils (e.g. stomata, xylem vessels)		Anticipation of the plane of new cellular divisions; formation of new cell-edge MTOC

control the orientation of the cellulose microfibrils, thereby determining the shape and morphology of the new growing cell [8, 9]. However, depending on the type of plant and tissue, other structures like e.g. the microfilaments may also participate in the morphogenesis of the cell [6].

Morphogenesis is also created via the planes of new cell divisions. Equal or unequal cell divisions are possible. The new division plane is anticipated by a "pre-prophase band" of microtubules which, however, disappears completely before the actual cell division becomes obvious. Since the new primary cell wall, which is formed after nuclear division, bisects the pre-prophase band area, both daughter cells inherit half of the pre-prophase band width. This points to an exceptional importance of the cell edges, which, due to this mechanism, have been inherited inevitably from the respective mother cells. Therefore, the "lattice of cell edges may be a sensory system used by the plant to perceive positional information and to translate it into actions in which microtubules participate" [5]. The cortical microtubules then should spread over the interior cellular surfaces, originating from the cell edges and controlling the growth of the secondary walls. The net of cortical microtubules, forming a fibrillar cytoskeleton, has been visualized by immunofluorescence [10–12].

II. Flagella, Cilia, and Other Microtubular Systems

As pointed out above, higher plants (conifers and angiosperms) do not contain centrioles as spindle MTOC's. Nor do higher plants possess the ability to form flagella or cilia in any phase of their life cycle. Flagella and cilia are only generated in lower plants (algae, mosses, ferns, certain gymnosperms) for the locomotion of unicells. Their MTOC is the "basal body" or the "blepharoplast", a short cylindrical organelle with a peculiar cross-section [1, 13]. The cross-section of the flagella shows 9 outer doublets and 2 central microtubules and a number of arms and spokes (Fig. 38). The arms are thought to participate in a sliding mechanism which is presumed to create the force needed for the flagellar beating [1, 13].

The flagella are a particularly useful system for the study of microtubules. For example, the 2 flagella of the unicellular green alga *Chlamydomonas reinhardii* can be sheared off by a mild acid treatment and will then regrow in about 90 min (Fig. 39). After a lag of 12–15 min, flagellar regrowth and the synthesis of new tubulin start simultaneously. The synthesis of tubulin and other flagellar proteins is preceded by an increase in the levels of the respective m–RNAs [14]. This system has proved very useful in studies concerning the elucidation of the mode of action of herbicides interfering with the microtubular system (p. 130, 134).

Another microtubular system is restricted to the animal kingdom and will be mentioned here only briefly. The axon of nerve cells is an aggregation of specialized microtubules, and also other signal-transducing systems appear to integrate microtubular structures. Transmission of stimuli in sensory cells has therefore been argued to occur along specialized microtubules [15].

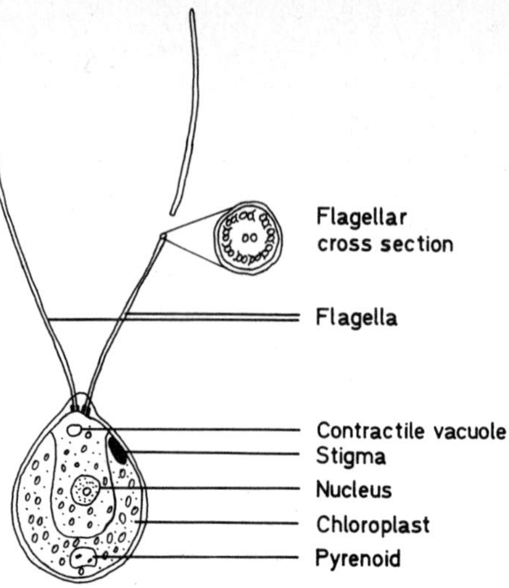

Fig. 38. *Chlamydomonas* as an example of a green algal flagellate cell, showing cellular organelles, two flagella, and a flagellar cross section [1]

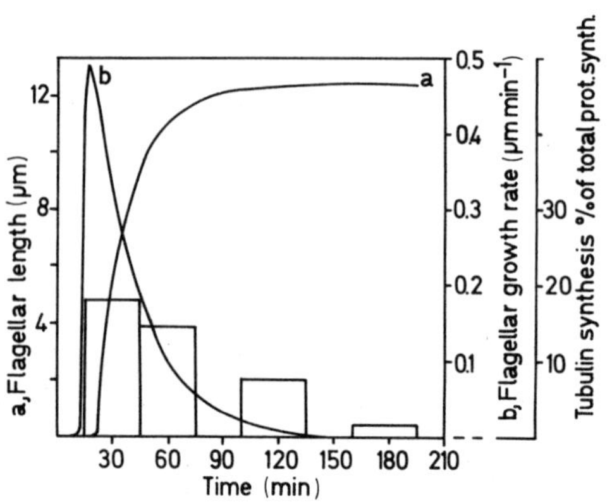

Fig. 39. Flagellar regeneration and simultaneous tubulin synthesis in the unicellular green alga *Chlamydomonas reinhardii* [2–4]. At zero time the algae have shed their flagella after a pH-shock treatment

III. Formation of Microtubules

As already mentioned, the microtubule-forming unit is the dimeric protein tubulin. The monomers α- and β-tubulin from plant cells differ from those in animal cells with respect to their molecular weight and also with respect to the colchicine-binding ability [6, 16]. Tubulin contains a number of different binding sites, which include one for colchicine, one for vinblastine, one for GTP, and one each for lateral and longitudinal binding needed during the formation of microtubules. Tubulin from animal sources can be polymerized to microtubules in vitro; this, however, has not yet been achieved with tubulin of plant or fungal origin. However, plant

Table 35. Polymerization of tubulin to microtubules

α-tubulin subunit / β-tubulin subunit →	Tubulin (dimer) (50 × 80 Å) →	[Rings, Double rings, Spirals, Stacks →	Sheets (Protofilaments)] →	Micro-tubules

Microtubule organizing center (MTOC)
Controlling factors: Ca^{2+}-concentration
Calmodulin
MAPs (MT associated proteins)
τ-protein: M ~ 75,000
HMW proteins: M ~ 200,000
GTP
Mg^{2+}

or fungal tubulin can be polymerized in heterologous systems together with animal (usually brain) tubulin [18].

These difficulties already point to differences among different microtubules. Moreover, microtubules also differ greatly in their in vivo stability, the order being nerve cell MTs > flagellar MTs > cytoplasmic MTs with decreasing stability. These differences in stability, but also differences in function, certainly reside at least in part in the MAPs (microtubule-associated proteins). The MAPs identified so far are the τ-factor (tau-factor) which enhances polymerization in vitro, and the HMW (high molecular weight) proteins, which however, are not required for polymerization [17]. In vivo and in vitro microtubular behaviour are certainly not strictly comparable, and many other factors will be included in the control of in vivo-microtubule formation. Some aspects of the microtubule polymerization process are summarized in Table 35. Rings, spirals, etc. are generated in vitro, but may not occur in vivo [19]. The sheets tend to curl and tend to eventually form hollow microtubule cylinders comprised of 13 protofilaments [7].

Among the factors mentioned in Table 35, the Ca^{2+}-concentration deserves special attention. Calcium has been recognized in recent years as an elementary bioregulatory agent. The Ca^{2+}-concentration in the cytosol may fluctuate between 0.1 and 10 μM, which is the concentration range where the protein calmodulin reversibly binds calcium. Calmodulin (or the similar proteins CdR, Troponin C) in turn regulates the activities of certain metabolic key enzymes which, as a common feature, are involved in the synthesis or breakdown of phosphate esters [20, 21]. These enzymes include NAD-kinase, cyclic nucleotide phosphodiesterase and phosphorylase b kinase [22]. Microtubules are unstable or may not polymerize at high Ca^{2+}-concentrations [20, 23]. Calcium also regulates the beating of flagella [24] and cilia [25]. The functions of Ca^{2+} appear to be similar in the plant and animal kingdom since calmodulin has been isolated from both plant and animal sources, and the calmodulin from both sources is interchangeable [26, 27].

The key factor for this type of regulation being the free Ca^{2+}-concentration in the cytoplasm, it is therefore of prime interest to know the Ca^{2+}-pumping systems

Table 36. Characterization of the Ca^{2+} transport systems in mitochondria and in plasma membrane vesicles from higher plants [28–30]

Parameter	Mitochondria	Plasma membrane vesicles
Ca^{2+}-uptake capacity	High	Low
(n mol/mg protein/15 min)	~200	~1
Ca^{2+}-saturation conc.	High	Low
Regulation by calmodulin	No	Yes
Activity variation among different tissues	High	Low
System dependencies	P_i, substrate	ATP

which, in turn, control the cytoplasmic Ca^{2+}-concentration. Table 36 compiles the characteristics of two such systems which have been studied in plant extracts. The mitochondrial system may have the function of a coarse control and the plasmalemma-derived system that of a fine control. Therefore, both systems may cooperate in maintaining the appropriate level of free cytoplasmic calcium.

IV. Morphological and Cytological Effects Induced by Herbicides and Other Compounds Interfering with Microtubular Systems

As might be suspected from a study of Table 34, the loss of microtubules should have dramatic consequences for the individual plant cell as well as for the morphogenesis of the whole plant. The loss of spindle microtubules should affect nuclear division and the separation of chromosomes whereas the morphogenesis of cells and tissues should be affected when cortical microtubules are lacking. The same effects may be observed when the polymerization of microtubules is not inhibited but when their normal subcellular organization is prevented. The latter appears to apply to carbamate herbicides (p. 133).

The morphological and cytological effects which may be attributed to an interference with microtubular systems are summarized in Table 37. However, a swelling of tissue does not necessarily involve interference with cortical microtubules [31, 32]. In addition to effects that are due to lacking or malfunctioning spindle and cortical microtubules, interference with the "nuclear-cage" microtubules in *Micrasterias denticulata* and with the flagellar microtubules in flagellates is also recorded in Table 37.

Besides the herbicides cited in Table 37, other compounds have been reported to induce similar effects. *Colchicine*, which binds to the tubulin dimer and thereby inhibits microtubule formation, apparently affects plant microtubules only at high concentrations [5, 80–84]. Other compounds which induce related morphological effects but for unknown primary reasons include lindane [85–87] and the plant growth inhibitor *ancymidol* [88, 89].

Table 37. Morphological and cytological effects of anti-microtubular herbicides on plant cells and tissues

Effect(s)	Organism(s)	Herbicide(s)	Reference
Consequences of lacking spindle microtubules:	Allium	Asulam	[33, 34]
	Vicia	Barban	[35]
	Vicia	Chlorpropham	[35]
	Vicia	Propham	[35]
Cell-cycle stop at metaphase; arrested metaphases accumulating; chromosome aberrations; polyploidization; bi- and polynuclear cells; giant nuclei	Triticum	Barban	[36]
	Allium	Chlorpropham	[37]
	Hordeum	Chlorpropham	[37]
	Glycine	Chlorpropham	[38]
	Haemanthus	Propham	[39, 40]
	Pisum	Propham	[41]
	Zea	Dinitramine	[42]
	Triticum	Dinitramine	[42]
	Gossypium	Trifluralin	[43]
	Haemanthus	Trifluralin	[44]
	Allium	Trifluralin	[45, 46]
	Vicia	Trifluralin	[47]
	Nicotiana	Trifluralin	[48]
	Allium	Pendimethalin	[49]
	Zea	Dinitroanilines	[50]
	Allium	S-2846	[51]
	Setaria	Chlorthal-dimethyl	[52]
	Avena	Pronamid	[53]
	Cucumis	Pronamid	[53]
	Zea	Sindone B	[42]
	Triticum	Terbutol	[42]
Consequences of lacking cortical microtubules:	Triticum	Barban	[36]
	Haemanthus	Propham	[54]
	Allium	Propham	[55]
	Mougeotia	Propham	[56]
Deranged cell-wall structures (vascular tissue, stomata); irregular cell morphology; swelling of cells; crushed xylem; decreased length/width ratio of elongating cells; root swelling in elongation zone; no or little secondary cell-wall thickening	Pediastrum	Propham	[57]
	Nicotiana	Benfluralin	[58]
	Zea	Oryzalin	[59]
	Chara	Trifluralin	[60]
	Glycine	Trifluralin	[61]
	Allium	Trifluralin	[62, 63]
	Triticum	Trifluralin	[62]
	Gossypium	Trifluralin	[63]
	Echinochloa	Trifluralin	[63]
	Carthamus	Trifluralin	[63]
	Beta	Trifluralin	[64]
	Oocystis	Trifluralin	[65]
	Oocystis	Oryzalin	[65]
	Oocystis	Amiprophos-methyl	[65]
	Triticum	Amiprophos-methyl	[66]
	Chlorella	Butamiphos	[67]
	Triticum	Isophos	[68]
	Avena	Chlorthal-dimethyl	[69]

Table 37 (continued)

Effect(s)	Organism(s)	Herbicide(s)	Reference
	Avena	Pronamid	[53]
	Cucumis	Pronamid	[53]
	Triticum	Pronamid	[70]
No post-mitotic nuclear migration	Micrasterias	Chlorpropham	[71, 72]
		Propham	[71, 72]
		Trifluralin	[72]
		Amiprophos-methyl	[73]
Flagellar or ciliar regeneration inhibited	Ochromonas	Chlorpropham	[74]
	Ochromonas	Propham	[74]
	Stentor	Propham	[75]
	Chlamydomonas	Oryzalin	[76]
	Chlamydomonas	Amiprophos-methyl	[76, 77]
	Chlamydomonas	Trifluralin	[78]
	Stentor	Trifluralin	[79]

V. Herbicide Interference with Microtubular Systems

Of the herbicides presently known to interfere with microtubular systems, two structural series have been known for a long time: dinitroanilines (Table 38) and carbamates (Table 39). Phosphoric amides constitute a rather new group of herbicides (Table 40). Their mode of action has, however, already been studied extensively. Table 41 compiles herbicides which, although structurally dissimilar to the aforementioned compounds, also interfere with the assembly or function of microtubules.

1. Dinitroanilines

After treatment with dinitroaniline herbicides, no microtubules can be detected in the roots [42, 90]. In a review, Parka et al. [91] have named 14 dinitroaniline herbicides which all appear to display the same primary mode of action, i.e. interaction with the microtubular system. They may, however, differ with respect to additional primary modes of action, depending on concentration. As for all the herbicides interfering with the microtubular system, gramineous species are most sensitive. The lowest active concentration was 6 nM in corn and 20 nM in soybean [92].

The molecular mode of action is not yet clear. Binding of *trifluralin* to a protein from *Chlamydomonas eugametos* flagellae, which exhibited the characteristics of tubulin, has been reported [92a]. Binding occurring at 5.8 μM was not competitive with colchicine and was claimed to be specific. The characterization of the binding protein is currently reinvestigated in more detail. However, other authors did not observe specific binding of dinitroaniline herbicides in plant extracts [93]. Binding of *trifluralin* to tubulin of animal origin has not been detected in one series of experiments, nor did *trifluralin* inhibit the in vitro polymerization of brain tubulin

Table 38. Dinitroaniline herbicides

Common name	Structure	Water sol. (mg l^{-1})
Benfluralin = benefin	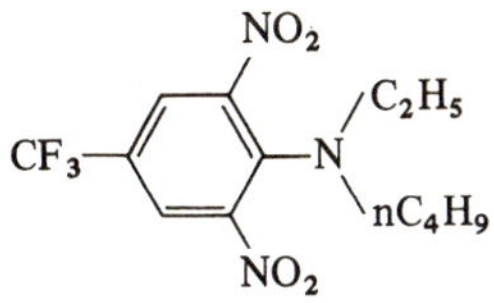	0.2
Dinitramine	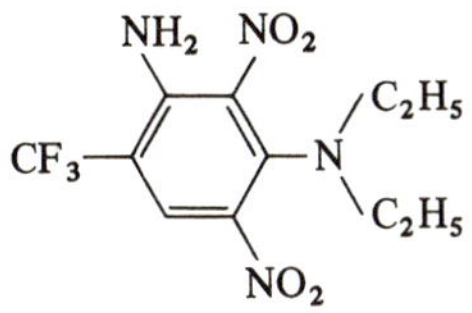	1.0
Fluchloralin	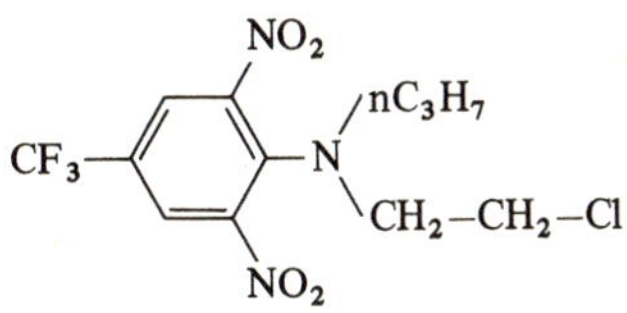	70
Nitralin	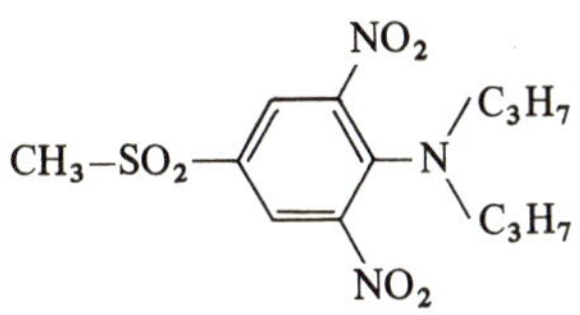	0.6
Oryzalin	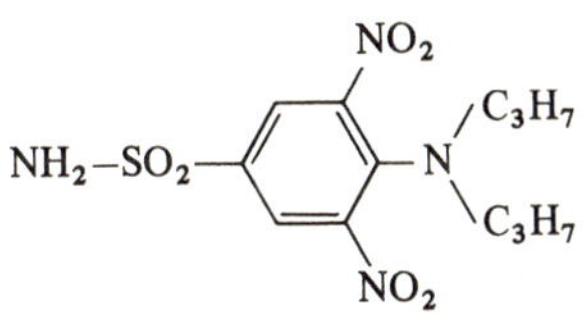	2.5
Pendimethalin = penoxalin	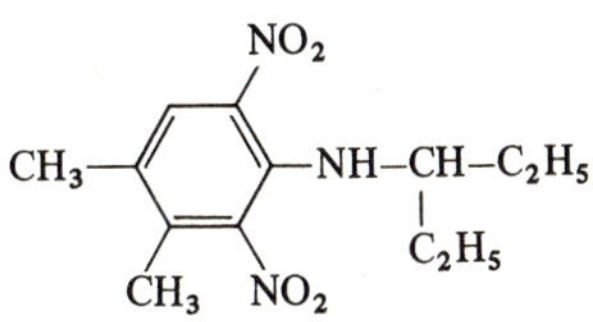	0.3
Trifluralin	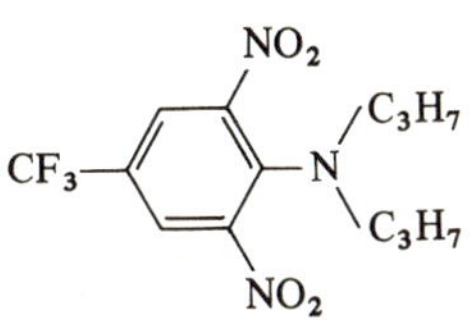	0.25

Table 39. Carbamate herbicides

Common name	Structure	Water sol. ($mg\ l^{-1}$)
Asulam	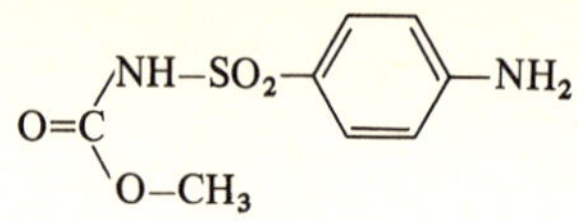	4,000
Barban	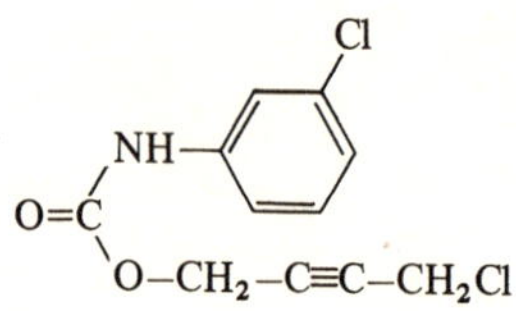	11
Chlorpropham =CIPC	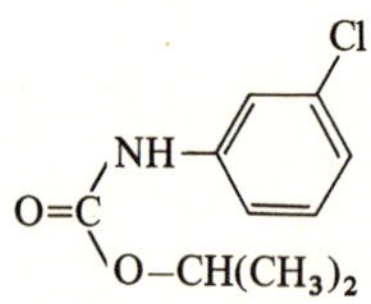	89
Propham = IPC	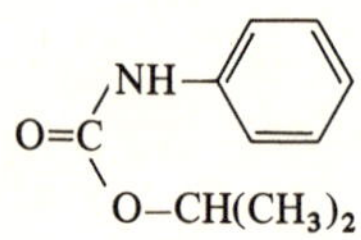	250

Table 40. Phosphoric amide herbicides

Common name	Structure	Water sol. ($mg\ l^{-1}$)
Amiprophos-methyl	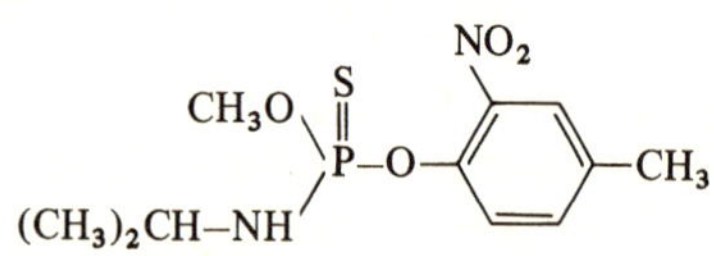	70
Isophos	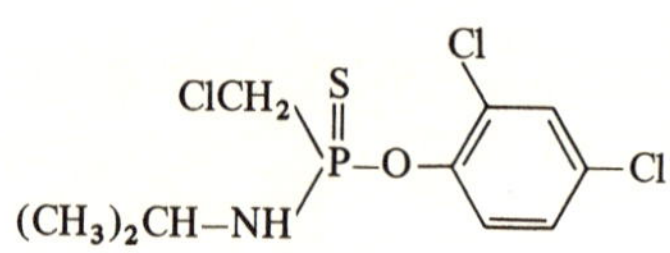	–
S-2846 = butamiphos	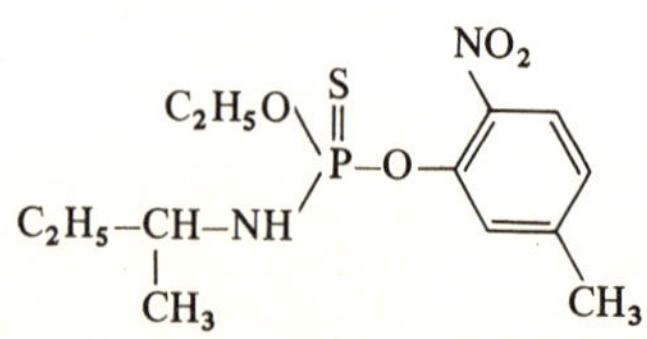	–

Table 41. Miscellaneous herbicides with antimitotic activity

Common name	Structure	Water sol. (mg l^{-1})
Chlorthal-dimethyl = DCPA	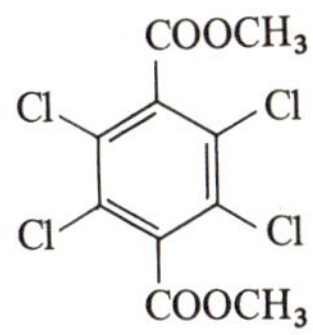	< 0.5
Pronamid = propyzamide	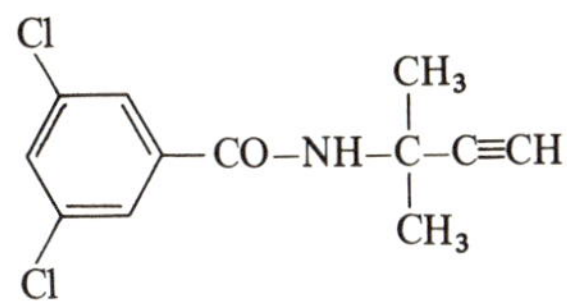	15
Sindone B	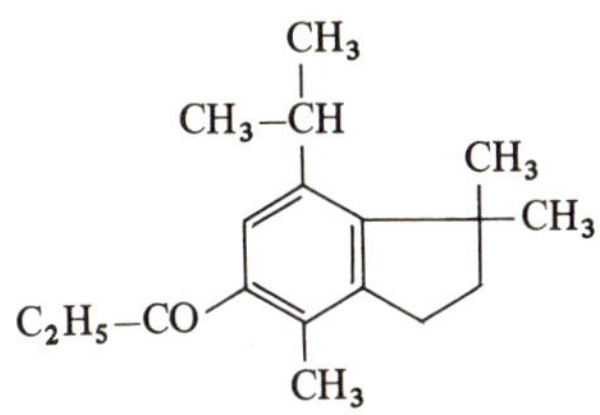	–
Terbutol	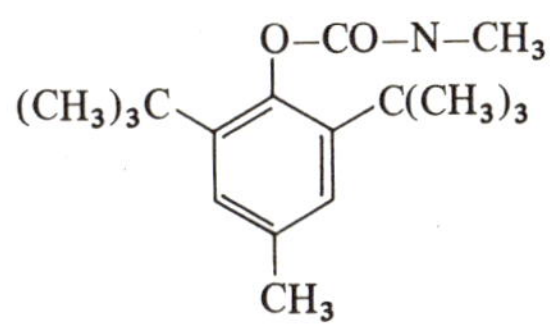	7

to microtubules [90]. Another report demonstrates partial inhibition of in vitro pig brain polymerization with *trifluralin* and *oryzalin* at 90% water-solubility concentration [94]. These conditions might not be relevant to the in vivo situation. *Trifluralin* did not inhibit the growth of monolayer chick lung cell cultures [92a].

Clearly, then, dinitroaniline herbicides at herbicidal concentrations only affect plant microtubular systems and do not interfere with tubulin or microtubules of animal origin.

2. Carbamates

In the presence of carbamate herbicides, microtubules are still found in the tissue, at least at low concentrations. However, these microtubules are no longer arranged in the usual ordered way [90, 95, 96]. It has therefore been inferred that carbamates interfere with the MTOC (Table 34) and disturb the directionality of a new microtubular assembly. Consequently, the microtubules can no longer fulfill their roles in cellular morphogenesis. Although binding of chlorpropham to brain tubulin did not occur and in vitro polymerization of brain tubulin was not inhibited [90,95], at high concentrations carbamates may also affect the MTOCs of animal cells [98–100]. This is e.g. indicated by abnormal MTOC multiplication and by multipolar spindles.

3. Phosphoric Amides

Studies on the molecular mode of action of these herbicides have been made with *amiprophos-methyl* and algae. In the case of *Oocystis* a comparison of several antimicrotubular herbicides has shown that, after treatment with *amiprophos-methyl*, *trifluralin*, or *oryzalin*, microtubules were virtually absent in the cell and cell-wall organization was abnormal [94]. The correlation between cellulose microfibril and cortical microtubule directionality has been extensively studied and confirmed for *Oocystis* using *colchicine*, *amiprophos-methyl*, and the protein-synthesis inhibitor *cycloheximide* [65]. *Amiprophos methyl* neither interferes with in vitro brain tubulin polymerization, even at high concentrations [76,94], nor affects animal tissue culture cells [65].

Amiprophos-methyl, like dinitroanilines, affects flagellar regeneration in *Chlamydomonas reinhardii* at μM concentrations [76, 100]. Simultaneously, the synthesis of new tubulin (Fig. 39) and one additional flagellar protein is inhibited [77]. Interestingly, the synthesis of other flagellar proteins, also induced by deflagellation, is not inhibited. The inhibition of tubulin synthesis appears to be caused by a lack of the specific m-RNA. The same is true when *amiprophos-methyl* is added after tubulin synthesis has already started, indicating that m-RNA in tubulin synthesis may be rapidly degraded in the presence of *amiprophos-methyl*. Alternatively, if the tubulin m-RNA turnover is very high, the herbicide may inhibit new tubulin m-RNA synthesis [100a].

4. Other Herbicides and Additional Compounds Interfering with Microtubules

The herbicides *sindone B* and *terbutol* have been reported to cause a lack of spindle microtubules in the roots of corn and wheat [101]. However, both compounds did not affect the cortical microtubules. *Pronamid*, in a study together with *trifluralin* and *chlorpropham*, was shown to inhibit the formation of microtubules in roots but to have no effect on in vitro polymerization of brain tubulin [90].

An interesting case are the benzimidazole fungicides which inhibit microtubule formation only in fungi [18]. These fungicides and/or their metabolite *MBC* (methylbenzimidazolyl carbamate) bind to tubulin from aspergillus in a competitive manner with colchicine. The resistance to this fungicide appears to be correlated with a loss of tubulin binding. However, some benzimidazole fungicides might act differently [102]. The related anthelminthic drug *mebendazole* inhibits in vitro brain tubulin polymerization [103].

From the information available it appears that herbicide and fungicide binding to brain tubulin in vitro and also the interference with in vitro brain tubulin polymerization systems should be interpreted with care. In particular, the in vitro polymerization of brain tubulin to microtubules occurs spontaneously, i.e. apparently without participation of MTOC. Specific proteins are present (Table 34), but other so far unknown factors may be important for other systems. Only where physiologically significant binding occurs to a tubulin site necessary for the polymerization process, as is the case with colchicine, may in vitro polymerization data be relevant. At present, our knowledge of plant microtubular systems at a molecular level is not sufficient to draw any firm conclusions.

5. Calcium Pumps

As mentioned before (Table 35), Ca^{2+} appears to play a major role in the regulation of the polymerization/depolymerization of microtubules. It has therefore been suggested that some of the antimicrotubular herbicides might affect microtubular formation via a deregulation of the cytoplasmic Ca^{2+}-level [76].

One attempt to explain the action of antimicrotubular herbicides on a Ca^{2+}-dependent molecular level is based on the possible interference with the Ca^{2+} pumping system in the outer mitochondrial membrane (Table 36). Inhibition has only been achieved for the mitochondrial but not for the microsomal system. The data listed in Table 42, obtained at 0.1 mM herbicide concentration, indicate only a limited specificity. Other herbicides, like e.g. *2,4,5-T* and *endothal*, were inactive. Some herbicides such as *ioxynil* may affect Ca^{2+}-uptake indirectly, at least at higher concentrations, because they interfere with the energy-transducing system. The most active herbicides were *amiprophos-methyl*, *trifluralin*, and *oryzalin* which had I_{50}-values of 6 μM, 0.7 μM, and 20 μM, respectively, and *ioxynil* which primarily inhibits mitochondrial energy conservation. In the case of *amiprophos-methyl* inhibition of Ca^{2+}-uptake at 10 μM herbicide concentration was additionally demonstrated [105]. At higher concentrations and with the other herbicides

Table 42. Effect of some herbicides known to interfere with mitosis on the Ca^{2+}-accumulation and oxidative phosphorylation in mitochondria from corn and squash [104]

Herbicide	Effect at 0.1 mM	
	Ca^{2+}-accumulation in mitochondria (% of control)	Oxidative phosphorylation
Amiprophos-methyl	0– 5	Little effect
Chlorpropham	9–32	Slight inhibition
Chlorthalmethyl	22–23	Slight uncoupling
Ioxynil	0– 3	Strong inhibition
Oryzalin	0	Strong uncoupling
Propham	47–71	Little effect
Trifluralin	0– 2	Inhibition

(Table 42) Ca^{2+} was shown to leak out of the mitochondria in the presence of the herbicides. Inhibition of Ca^{2+}-uptake has also been demonstrated to occur with *trifluralin* and *amiprophos-methyl* in *Chlamydomonas* [76].

VI. Metabolic Effects of Antimitotic Herbicides

The antimitotic herbicides dealt with in this chapter of course affect other paths of plant metabolism as a consequence of their primary interference with microtubular systems. Among the metabolic pathways affected indirectly, nucleic acid and protein synthesis may be the most important. However, additional modes of interference with plant metabolism have also been observed, e.g. the inhibition of photosynthesis by dinitroanilines.

1. Dinitroaniline Herbicides

Inhibition of photosynthesis by *trifluralin* in isolated chloroplasts has been investigated by Robinson et al. [106]. They found a pI_{50} value of 4.6 and reported the inhibition site to be located between the photosystems I and II. More precisely, several dinitroaniline herbicides affect the electron transport in isolated chloroplasts at the reducing side of the plastoquinone pool, i.e. they act as DBMIB-analogues (p. 24) [107, 108]. Alsop et al. [109] reported inhibition and uncoupling of coupled photosynthetic electron flow at an *oryzalin* concentration ranging from 10 to 100 μM.

Respiration and oxidative phosphorylation are also inhibited by dinitroaniline herbicides, depending on concentration. *Trifluralin* and *nitralin* exhibit inhibitory effects above 100 μM whereas *oryzalin* is active at somewhat lower concentrations [110, 111]. Moreland et al. [112, 113] studied the influence of 12 dinitroanilines on mitochondrial and chloroplast activities in vivo and in vitro and found inhibitory activities in the 0.1–1.0 mM and 0.01–0.1 mM concentration range, respectively. *Oryzalin* caused respiratory inhibition whereas *trifluralin* induced respiratory uncoupling in vivo. In a more recent study, all 12 dinitroanilines where reported to inhibit all of the tested partial reactions in mung bean mitochondria, albeit to a different extent [114, 115]. Generally, the differences were quantitative rather than qualitative. It was therefore postulated that inhibitions caused by dinitroaniline herbicides could not be attributed to specific molecular sites of interaction but that membrane fluidity and membrane permeability were changed instead. The molecular basis for this effect may be seen in the partitioning of the relatively lipophilic herbicide molecules into the lipophilic phase of the membrane. The increase in permeability seen after treatment of soybean roots with *dinitramine* may have a similar cause [116].

The concept of membrane fluidity changes, also termed "membrane perturbations" [115], caused by an "unspecific" partitioning of lipophilic herbicide molecules into the membranes might be of general importance. Consequences for the membrane functions could possibly include partial uncoupling of energy-conserving systems in mitochondria and chloroplasts, membrane permeability changes towards ions and other molecules, and changed responses of the membrane in a number of stress situations. The herbicidal concentrations needed for this type of interference are in the 0.01–1 mM range, i.e. well above those generally required for the herbicidal action. However, cases might exist where this type of interference could be of some relevance in vivo.

Physiological and metabolic effects which might occur as a consequence of the inhibition of cell divisions include a decreased chromatin activity [117] and inhibition of nucleic acid and protein synthesis. Water solubilities and I_{50}-values for the inhibition of the growth of tobacco calluses (1.6×10^{-8}–9×10^{-6} M) for 11 dinitroanilines have been reported by Huffmann et al. [119].

It may also be worth mentioning that the herbicidal activity can be antagonized by cysteine [120] and by lipophilic compounds like tocopherol, palmitate, vitamin K_1, and others [121].

2. Carbamates

An early hypothesis formulated that carbamate herbicides were "inhibitors of changes in metabolic state" [122]. This "inhibition of changes in metabolic state" can be seen as a consequence of the inhibition of cell divisions which eventually leads to a slow-down of many cellular metabolic pathways. Inhibitions have primarily been observed for RNA and protein syntheses, e.g. after bracken treatment with *asulam* [123], after treatment of oats with *barban* [124, 125], and after treatment of tobacco callus with *chlorpropham* and *propham* [126]. As might be expected, shoot formation in tobacco callus cultures was also inhibited [127]. A particular system where the herbicidal influence on the induction of new enzyme synthesis can be measured utilizes the gibberellic acid induced α-amylase formation in barley half-seeds [122, 128]. Potatoes kept at either 3 or 11 C and treated with *chlorpropham* or *propham* have also shown inhibitions of protein synthesis and processes related to germination [129, 130]. However, when studying the photocontrol of respiration in germinating *Lactuca*, Woodstock et al. [131] found no influence of *chlorpropham* on this process. Germination of *Lactuca* was delayed in the presence of *chlorpropham*. The inhibitory effects of *chlorpropham* on the growth of chlorella were also thoroughly studied [132]. Whereas respiration was not affected at more than 10 times the I_{50}-concentration for growth inhibition, the DNA-content was found to increase in the inhibited cells. The I_{50}-value was 4 μM in the light and 1.3 μM in the dark.

Respiration appears not to be affected directly by carbamate herbicides. However, this may be the case at higher concentrations; thus, e.g. *chlorpropham* at >0.1 mM greatly decreases the level of ATP in soybean hypocotyls [133]. *Chlorpropham* is metabolized to 2-OH- and/or 4-OH-chlorpropham, of which only the former inhibits oxidative phosphorylation [134–137]. However, sensitive cucumber metabolized chlorpropham to the 4-OH-metabolite whereas resistant corn yielded the 2-OH-compound. Sensitive oats take up more herbicide in the case of *barban* [125] and further metabolize the 4-OH-chlorpropham to the S-cysteinyl derivative [138–140].

The NH-CO part of the carbamate molecule is also present in urea herbicides and other inhibitors of photosystem II. This coincidence has led researchers repeatedly to measure the inhibition of photosynthesis in the presence of carbamate herbicides in isolated chloroplasts. The pI_{50}-value was between 3.4 and 3.8 for propham and between 3.8 and 4.0 for chlorpropham [141–143]. However, these inhibition constants are very high as compared with other photosystem II inhibitors (Tables 4–15), and no in vivo inhibition of photosynthesis was found [144].

Thiolcarbamate herbicides obviously exhibit a completely different mode of action (p. 142), since they do not show the pronounced morphological and cytological effects peculiar to the antimicrotubular herbicides. On the other side, *chlorpropham* did not influence the epicuticular wax formation [145], an effect commonly displayed by thiolcarbamate herbicides.

Asulam has recently been reported to be a competitor of p-aminobenzoic acid and of folic acid at concentrations of 10 and 2 μM in carrot cell suspension cultures [146]. This observation was confirmed for the roots of intact plants, suggesting an interference of *asulam* with folic acid synthesis [146a]. Also, *asulam*-resistant cell lines could be obtained with celery-cell suspension cultures [147].

References for F

1. Mohri, H.: Biochim. Biophys. Acta *456*, 85 (1976)
2. Weeks, D.P., Collis, P.S., Gealt, M.A.: Nature *268*, 667 (1977)
3. Weeks, D.P., Collis, P.S.: Cell *9*, 15 (1976)
4. Weeks, D.P., Collis, P.S.: Developmental Biol. *69*, 400 (1979)
5. Gunning, B.E.S., Hardham, A.R.: Endeavour *3*, 112 (1979)
6. Filner, P., Yadav, N.S.: Role of microtubules in intracellular movements. In: Encyclopedia in plant physiol., New Series, Vol. 7. Haupt, W., Feinleib, M.E. (eds.), p. 95. Berlin, Heidelberg, New York: Springer 1979
7. Dustin, P.: Scientific Amer. *243*, 66 (1980)
8. Hepler, P.K.: Plant microtubules. In: Plant biochemistry. Bonner, J., Varner, J.E. (eds.), p. 147. New York, San Francisco, London: Academic Press 1976[3]
9. Hardham, A.R., Green, P.B., Lang, J.M.: Planta *149*, 181 (1980)
10. Lloyd, C.W. et al.: Nature *279*, 239 (1979)
11. Lloyd, C.W. et al.: Planta *147*, 500 (1980)
12. Powell, A.J. et al.: Plant Sci. Lett. *18*, 401 (1980)
13. Satir, P., Ojakian, G.K.: Plant cilia. In: Encyclopedia in plant physiol, New Series, Vol. 7. Haupt, W., Feinleib, M.E. (eds.), p. 224. Berlin, Heidelberg, New York: Springer 1979
14. Lefebvre, P.A. et al.: Cell *20*, 469 (1980)
15. Atema, J.: Stimulus transmission along microtubules in sensory cells: an hypothesis. In: Microtubules and microtubule inhibitors. Borgers, M. Brabander, M. (eds.), p. 247. Amsterdam: North-Holland Publ. Comp. 1975
16. Rubin, R.R., Cousins, E.H.: Phytochem. *15*, 1837 (1976)
17. Schmitt, H., Josephs, R., Reisler, E.: Nature *265*, 653 (1977)
18. Davidse, L.C.: Antimitotic activity of methyl benzimidazol-2 yl carbamate in fungi and its binding to cellular protein. In: Microtubules and microtubule inhibitors. Borgers, M., Brabander, M. (eds.), p. 483. Amsterdam: North-Holland Publ. Comp. 1975
19. Burns, R.: Nature *273*, 709 (1978)
20. Geisow, M.J.: Nature *276*, 211 (1978)
21. Means, A.R., Dedman, J.R.: Nature *285*, 73 (1980)
22. Anderson, J.M., Cormier, M.J.: Biochem. Biophys. Res. Comm. *84*, 595 (1978)
23. Marcum, J.M. et al.: Proc. Natl. Acad. Sci. U.S.A. *75*, 3771 (1978)

24. Schmidt, J.A., Eckert, R.: Nature *262*, 713 (1976)
25. Verdugo, P.: Nature *283*, 764 (1980)
26. van Eldik, L.J. et al.: Proc. Natl. Acad. Sci. U.S.A. *77*, 1912 (1980)
27. Dieter, P., Marmé, D.: Cell Calcium *1*, 279 (1980)
28. Gross, J., Marmé, D.: Proc. Natl. Acad. Sci. U.S.A. *75*, 1232 (1978)
29. Dieter, P., Marmé, D.: FEBS Lett. *125*, 245 (1981)
30. Dieter, P., Marmé, D.: Planta *150*, 1 (1980)
31. Itoh, T.: Plant Cell Physiol. *17*, 367 (1976)
32. Itoh, T.: Plant Cell Physiol. *17*, 385 (1976)
33. Sterrett, R.B., Fretz, T.A.: Hort. Sci. *9*, 33 (1974)
34. Sterrett, R.B., Fretz, T.A.: Hort. Sci. *10*, 161 (1975)
35. Mann, J.D., Storey, W.B.: Cytologia *31*, 203 (1966)
36. Burström, H.G.: Physiol. Plant. *21*, 1137 (1968)
37. Nasta, A., Günther, E.: Biol. Zentralbl. *92*, 37 (1973)
38. Davis, D.G. et al.: Physiol. Plant. *40*, 15 (1977)
39. Hepler, P.K., Jackson, W.T.: J. Cell Sci. *5*, 727 (1969)
40. Hepler, P.K., Jackson, W.T.: J. Cell Sci. *5*, 745 (1969)
41. Rost, T.R., Bayer, D.E.: Weed Sci. *24*, 81 (1976)
42. Bartels, P.G., Watson, C.: Abstr. Weed Sci. Soc. Amer. Meet., no. 163 (1976)
43. Hess, D., Bayer, D.: J. of Cell Sci. *15*, 429 (1974)
44. Jackson, W.T., Stetler, D.A.: Can. J. Bot. *51*, 1513 (1973)
45. Delcourt, A., Deysson, G.: Cytologia *41*, 75 (1976)
46. Lignowski, E.M., Scott, E.G.: Weed Sci. *20*, 267 (1972)
47. Babak, M.: Weed Abstr. *27*, no. 2702 (1978)
48. Young, L.W., Camper, N.D.: Pestic. Biochem. Physiol. *12*, 117 (1979)
49. Beuret, E.: Weed Res. *20*, 83 (1980)
50. Upadhyaya, M.K., Noodén, L.D.: Plant Cell Physiol. *18*, 1319 (1977)
51. Sumida, S., Ueda, M.: Plant Cell Physiol. *17*, 1351 (1976)
52. Chang, C.T., Smith, D.: Weed Sci. *20*, 220 (1972)
53. Carlson, W.C., Lignowski, E.M., Hopen, H.J.: Weed Sci. *23*, 155 (1975)
54. Sanger, Jacksson: J. Cell Sci. *8*, 303 (1971)
55. Palevitz, B.A., Hepler, P.K.: Planta *132*, 71 (1976)
56. Marchant, H.J., Hines, E.R.: Planta *146*, 41 (1979)
57. Marchant, H.J.: Protoplasma *98*, 1 (1979)
58. Nel, P.C., Voges, J.H.: Agroplantae *9*, 111 (1977)
59. Upadhyaya, M.K., Nooden, L.D.: Plant Cell Physiol. *19*, 133 (1978)
60. Hess, F.D., Bayer, D.E.: Abstr. Weed Sci. Soc. Amer. Meet., no. 160 (1976)
61. Kust, G.A., Struckmeyer, B.E.: Weed Sci. *19*, 147 (1971)
62. Lignowski, E.M., Scott, E.G.: Plant Cell Physiol. *12*, 701 (1971)
63. Bayer, D.E. et al.: Amer. J. Bot. *54*, 945 (1967)
64. Stevenson, D.W., Norris, R.F.: Amer. J. Bot. *61*, 11 (1974)
65. Quader, H., Wagenbreth, I., Robinson, D.G.: Cytobiol. *18*, 39 (1978)
66. Draber, W., Fedtke, C.: Herbicide interaction with plant biochemical systems. In: Adv. Pestic. Sci., Part 3. Geissbühler, H. (ed.), p. 475. Oxford, New York: Pergamon Press 1979
67. Sumida, S., Ueda, M.: Chlorella as a model system to study herbicidal mode of action and its application to a new herbicide, 0-ethyl–0–(3-methyl-6-nitrophenyl)–N-sec-butyl phosphorothioamidate (S–2846). In: Mechanism of pesticide action, ACS Symp. Ser. 2, p. 156. Washington: Amer. Chem. Soc. 1974
68. Luzhnova, M.I. et al.: Mode of action of isophos-3. In: Proc. fourth Int. Congr. Pestic. Chem. (IUPAC), no. IV-430, Zurich 1978
69. Shabany, B., Anderson, J.L.: Weed Res. *12*, 164 (1972)
70. Desai, P.D.: Weed Abstr. *23*, no. 1630 (1974)
71. Kiermayer, O., Hepler, P.K.: Naturwiss. *57*, 252 (1970)
72. Kiermayer, O.: Protoplasma *75*, 421 (1972)
73. Kiermayer, O., Fedtke, C.: Protoplasma *92*, 163 (1977)
74. Brown, D.L., Bouck, G.B.: J. Cell Biol. *61*, 514 (1974)
75. Banerjee, S., Margulis, L.: Nature *224*, 180 (1969)

76. Quader, H., Filner, P.: Eur. J. Cell Biol. *21*, 301 (1980)
77. Collis, P.S., Weeks, D.P.: Science *202*, 440 (1978)
78. Hess, F.D.: Exptl. Cell Res. *119*, 99 (1979)
79. Margulis, L., Banerjee, S., Kelleher, J.K.: Assay for antitubulin drugs in live cells: oral regeneration in Stentor coeruleus. In: Microtubules and microtubule inhibitors. Borgers, M., de Brabander, M. (ed.), p. 453. Amsterdam, Oxford: North Holland Publ. Comp. 1975
80. Hogetsu, T., Shibaoka, H.: Planta *140*, 15 (1978)
81. Hammerschlag, D.R.H., McCully, M.E.: Plant Sci. Lett. *19*, 151 (1980)
82. Hart, J.W., Sabnis, D.D.: J. Exp. Bot. *27*, 1353 (1976)
83. Hart, J.W., Sabnis, D.D.: Phytochem. *15*, 1897 (1976)
84. Hart, J.W., Sabnis, D.D.: Current Adv. Plant Sci. *26*, 1095 (1976)
85. Charnetski, W.A., Lichtenstein, E.P., Evert, R.F.: Can. J. Bot. *51*, 2111 (1973)
86. Puiseux-Dao, S. et al.: Protoplasma *91*, 325 (1977)
87. Lazår-Keul, G., Cračinn, C., Soran, V.: Biochem. Physiol. Pflanzen *172*, 191 (1978)
88. Montague, M.J.: Plant Physiol. *56*, 167 (1975)
89. Shive, J.B., Sisler, H.D.: Plant Physiol. *57*, 640 (1976)
90. Bartels, P.G., Hilton, J.L.: Pestic. Biochem. Physiol. *3*, 462 (1973)
91. Parka, S.J., Soper, O.F.: Weed Sci. *25*, 79 (1977)
92. Upadhyaya, M.K., Nooden, L.D.: Plant Physiol. Suppl. *61*, no. 820 (1978)
92a. Hess, F.D., Bayer, D.E.: J. Cell Sci. *24*, 351 (1977)
93. Upadhyaya, M.K., Nooden, L.D.: Plant Physiol. *66*, 1048 (1980)
94. Robinson, D.G., Herzog, W.: Cytobiologia *15*, 463 (1977)
95. Coss, R.A., Pickett-Heaps, J.D.: J. of Cell Biol. *63*, 84 (1974)
96. Brower, D.L., Hepler, P.K.: Protoplasma *87*, 91 (1976)
97. Timson, J.: Pestic Sci. *1*, 191 (1970)
98. Lloyd, C.: Nature *277*, 515 (1979)
99. Oliver, J.M., Krawiec, J.A., Berlin, R.D.: Exp. Cell Res. *116*, 229 (1978)
100. Quader, H., Cherniade, J., Filner, P.: Plant Physiol. Suppl. *59*, no. 102 (1977)
100a. Minami, S.A. et al.: Cell, in press
101. Bartels, P.G., Watson, C.: Abstr. Weed Sci. Soc. Amer. Meet., no. 163 (1976)
102. Kumari, L., Decallonne, Meyer, J.A.: Pestic. Biochem. Physiol. *7*, 273 (1977)
103. Laclette, J.P., Guerra, G., Zetina, C.: Biochem. Biophys. Res. Comm. *92*, 417 (1980)
104. Hertel, C. et al.: Planta *149*, 336 (1980)
105. Hertel, C., Marmé, D.: personal communication
106. Robinson, S.J. et al.: Plant Physiol. *60*, 840 (1977)
107. Trebst, A., Draber, W.: Structure activity correlations of recent herbicides in photosynthetic reactions. In: Advances in Pestic. Science, Part 2. Geissbühler, H. (ed.), p. 223. Oxford, New York: Pergamon Press 1979
108. Trebst, A.: Inhibitors in electron flow: tools for the functional and structural localization of carriers and energy conservation sites. In: Methods in enzymology, Vol. 69. San Pietro, A. (ed.), p. 675. New York, London, Toronto, Sydney, San Francisco: Academic Press 1980
109. Alsop, W.R., Moreland, D.E.: Pestic. Biochem. Physiol. *5*, 163 (1975)
110. Negi, N.S.: Weed Sci. *16*, 83 (1968)
111. Wang, B., Grooms, S., Frans, R.E.: Weed Sci. *22*, 64 (1974)
112. Moreland, D.E., Farmer, F.S., Hussey, G.G.: Pestic. Biochem. Physiol. *2*, 342 (1972)
113. Moreland, D.E., Farmer, F.S., Hussey, G.G.: Pestic. Biochem. Physiol. *2*, 354 (1972)
114. Moreland, D.E., Huber, S.C.: Pestic. Biochem. Physiol. *11*, 247 (1979)
115. Moreland, D.E., Huber, S.C.: Fluidity and permeability changes induced in the inner mitochondrial membrane by herbicides. In: Plant mitochondria, developments in plant biology, Vol. 1, p. 181. Marseille: Symposium on Plant Mitochondria Proc. 1978
116. Travis, R.L., Woods, W.G.: Plant Physiol. *60*, 54 (1977)
117. Penner, D., Early, R.W.: Weed Sci. *20*, 364 (1972)
118. Schultz, D.P., Fuderburk, Jr., H.H., Negi, N.S.: Plant Physiol. *43*, 265 (1968)
119. Huffmann, J.B., Camper, N.D.: Weed Sci. *26*, 527 (1978)
120. Shahied, S.J., Giddens, J.: Agron. J. *62*, 306 (1970)
121. Hilton, J.L. et al.: Plant Physiol. *48*, 171 (1971)
122. Mann, J.D. et al.: Biochim. Biophys. Acta *138*, 133 (1967)

123. Veerasekaran, P., Kirkwood, R.C.: Weed Res. *17*, 85 (1977)
124. Kobayashi, K., Ishizuka, K.: Weed Sci. *22*, 131 (1974)
125. Kobayashi, K., Ishizuka, K.: J. Pestic. Sci. *2*, 59 (1977)
126. Keitt, G.W.: Physiol. Plant. *20*, 1076 (1967)
127. Yung, K.H., Palmer, C.E., Smith, O.E.: Plant Cell Physiol. *11*, 677 (1970)
128. Devlin, R.M., Cunningham, R.P.: Weed Res. *10*, 316 (1970)
129. Nowak, J.: Z. Pflanzenphysiol. *81*, 113 (1977)
130. Nowak, J.: Z. Pflanzenphysiol. *81*, 125 (1977)
131. Woodstock, L.W., Toole, V.K.: Plant Cell Physiol. *18*, 1 (1977)
132. Sumida, S., Yoshida, R., Ueda, M.: Plant Cell Physiol. *18*, 9 (1977)
133. Gruenhagen, R.D., Moreland, D.E.: Weeds *19*, 319 (1971)
134. Still, G.G., Mansager, E.R.: Phytochem. *11*, 515 (1972)
135. Still, G.G., Mansager, E.R.: Pestic. Biochem. Physiol. *3*, 289 (1973)
136. Shimabukuro, R.H., Walsh, W.C., Hoerauf, R.A.: Pestic. Biochem. Physiol. *6*, 115 (1976)
137. Rusness, D.G., Still, G.G.: Pestic. Biochem. Physiol. *4*, 24 (1974)
138. Still, G.G., Rusness, D.G.: Pestic. Biochem. Physiol. *7*, 210 (1977)
139. Rusness, D.G., Still, G.G.: Pestic. Biochem. Physiol. *7*, 220 (1977)
140. Rusness, D.G., Still, G.G.: Pestic. Biochem. Physiol. *7*, 232 (1977)
141. Wakamori, S.: Agr. Biol. Chem. *37*, 307 (1973)
142. Nishimura, K. et al.: Agr. Biol. Chem. *39*, 867 (1975)
143. Moreland, D.E.: Inhibitors of chloroplast electron transport: structure-activity relations. In: Progress in photosynthesis research, Vol. 3. Metzner, H. (ed.), p. 1693. Tübingen: Laupp Press 1969
144. Chow, P.N.P.: Can. J. Plant Sci. *56*, 429 (1976)
145. Still, G.G., Davis, D.G., Zander, G.L.: Plant Physiol. *46*, 307 (1970)
146. Killmer, J.L., Widholm, J.M., Slife, F.W.: Plant Sci. Lett. *19*, 203 (1980)
146a. Stephen, N.H., Cook, G.T., Duncan, H.J.: Ann. Appl. Biol. *96*, 227 (1980)
147. Merrick, M.M.A., Collin, H.A.: Plant Sci. Lett. *20*, 291 (1981)

G. Lipid Metabolism

I. Inhibition of Germination by Thiolcarbamate Herbicides

Thiolcarbamate herbicides are inhibitors of germination (Table 43). Like carbamate herbicides, they primarily affect gramineous species and display little phytotoxicity in dicotyledoneous species. However, in contrast to the carbamates (p. 133), they do not lead to swollen root tissue. On the contrary, shoots have been found to be more sensitive to thiolcarbamate herbicides than roots [1]. Although cell swelling occurs and aberrant mitoses have been observed, the primary effect of these herbicides clearly is on cell elongation rather than on cell division [2, 3].

II. Interference of Thiolcarbamates with Fatty Acid and Wax Formation

Early observations indicated that after treatment with thiolcarbamate herbicides the deposition of epicuticular waxes on the leaf surface was altered (for details and references see Table 44). The observed changes did not always include a decrease of the total amount of waxes deposited on the leaf surface, but an altered wax structure and/or cuticle thickness appears to be a general phenomenon after treatment with thiolcarbamates. As a consequence, the penetration of other herbicides applied to the leaf surface is facilitated. Also, in cases where the transpiration of the treated plant is increased, xylem-mobile herbicides may be taken up more extensively via the roots.

In parallel with the affected epicuticular wax formation, the extent of the fatty acid synthesis for phospholipids, monogalactosyl diglycerides (MGDG) and digalactosyl diglycerides (DGDG) is also decreased. In particular, the fraction of long-chain fatty acids and related alcohols and alkanes is reduced ($\geqq$C 28). The total amount of phospholipids is also reduced, but this may be seen as a consequence of a reduced supply of the long-chain fatty acids [12]. The biochemical basis for these effects may be an interference of thiolcarbamate herbicides with the synthesis of long-chain fatty acids. This is exactly what has been found in the investigations on the incorporation of ^{14}C-acetate and -malonate into the lipid fractions of seedling tissue, isolated chloroplasts and potato tuber slices (Table 44).

A recent review has appeared which discusses herbicidal effects on plant membrane lipids [41].

Table 43. Thiolcarbamate herbicides and some other herbicides exhibiting a strong inhibitory action on lipid synthesis

Common name	Structure	Water sol. ($mg\ l^{-1}$)
Benthiocarb	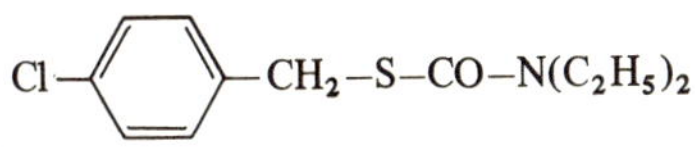	30
Butylate	$C_2H_5-S-CO-N[CH_2-CH(CH_3)_2]_2$	45
CABAC 39	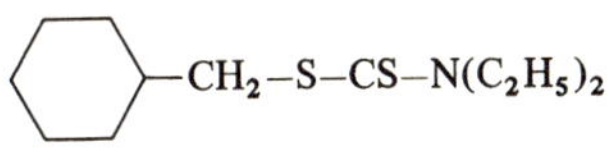	–
CDEC = sulfallate	$CH_2=CCl-CH_2-S-CS-N(C_2H_5)_2$	100
Cycloate	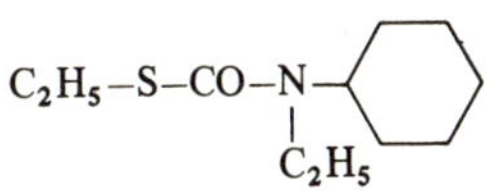	85
Di-allate	$HCCl=CCl-CH_2-S-CO-N[CH(CH_3)_2]_2$	14
EPTC	$C_2H_5-S-CO-N(CH_2-CH_2-CH_3)_2$	365
Molinate	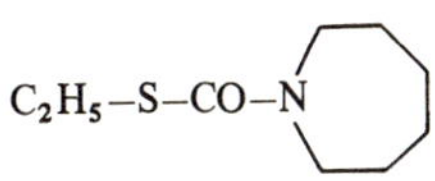	800
Pebulate	$CH_3-CH_2-CH_2-S-CO-N(C_2H_5)-nC_4H_9$	60
Tri-allate	$Cl_2C=CCl-CH_2-S-CO-N[CH(CH_3)_2]_2$	4
Vernolate	$CH_3-CH_2-CH_2-S-CO-N(CH_2-CH_2-CH_3)_2$	107
Dalapon	$CH_3-CCl_2-COONa$	($502\ g\ l^{-1}$)
Dinoben	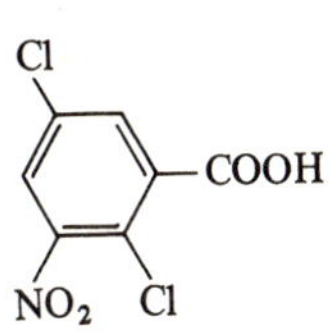	–
TCA	$Cl_3C-COONa$	($10\ kg\ l^{-1}$)

Table 44. Influence of thiolcarbamate herbicides on fatty acid and was formation

Herbicide(s)	Reference	Organism(s)	Effect(s)
CDEC, EPTC	[4]	Pea	Epicuticular wax formation affected: thinner cuticula; total amount not always reduced; changes in wax fine structure; uptake and accumulation of other herbicides increased; transpiration not always increased
EPTC	[5]	Cassia	
Di-allate	[6]	Pea	
Di-allate	[10]	Cassia	
EPTC	[7]	Cabbage	
EPTC	[8]	Wheat	
EPTC	[9]	Wheat	
EPTC	[11]	Cabbage	
EPTC	[5]	Cassia	Decrease of fatty acids in phospholipid, MGDG, and DGDG fractions; decrease of long-chain carbon compounds ($\geq$C 28)
EPTC	[12]	Cassia	
EPTC	[13]	Cassia	
Multiple field applications	[14]	Leaf crop plants	Minor changes in fatty acid quantity and quality
EPTC, diallate, CDEC	[15]	Pea, safflower	Inhibition of incorporation of ^{14}C-acetate and -mevalonate into long-chain fatty acids ($\geq$C 20); total incorporation not affected
Benthiocarb, EPTC	[16]	Chlorella	
CDEC, di-allate, EPTC	[17]	Pea	
EPTC, di-allate	[18]	Spinach chloroplasts	
Butylate, pebulate, vernolate	[19]	Spinach chloroplasts	
EPTC, di-allate, tri-allate	[20]	Potato	

III. Possible Mode of Herbicidal Action

It is at present hard to see how the inhibition of the synthesis of some very long-chain fatty acids may eventually be connected with herbicidal action. Possibly, the lack of certain specific and vital long-chain fatty acids, alcohols or alkanes may suppress the seedling growth to an extent which results in its eventual inability to grow sufficiently large for survival.

An additional aspect has recently been added by Wilkinson et al. [21] who observed an inhibition of the formation of gibberellic acid by 0.125 $mg\,l^{-1}$ *EPTC*. Also, the metabolism of kaurene to kaurenoic acid was reduced. A sufficiently large decrease of the plant hormone gibberellic acid may well contribute to the activity of thiolcarbamate herbicides as germination inhibitors. Interestingly, GA_3 has been found to antagonize the inhibitory action of *molinate* [22] and of *EPTC* [23]. The auxin herbicide *2,4-D* also antagonizes the inhibitory action of *EPTC* [24, 25]. In this case, the antagonism has been suggested to arise from the opposite actions of *EPTC* and *2,4-D* on the rate of RNA-synthesis. To sum up, the information available does not yet sufficiently explain the herbicidal and germination-inhibiting activities of thiolcarbamate herbicides. An inhibition of several (related?) steps in the synthesis of complex lipids may be inferred. Fatty acid chain elongation is certainly one sensitive biosynthetic pathway. Wilkinson et al. [21] suggest the inhibition of mixed function oxidases in the synthesis of gibberellic acid. Similar or related enzymatic mechanisms for the synthesis of other complex lipids might be affected similarly.

Thiolcarbamate herbicides may not be active as such but only after their conversion into the respective sulfoxides [26]. The sulfoxides easily react with mercapto groups, and the molecular mode of herbicide interference may therefore involve the reaction of the respective sulfoxide with a critical SH-group on the target enzyme complex [27]. S-chloroallylthiolcarbamates are further metabolized to the respective chloroacroleines [28–30]. A decrease of the total SH-content in tissue treated with a thiolcarbamate has been detected together with an increase in the activity of peroxidases [31]. Antidotes against the herbicidal activity of thiolcarbamates act via an increase of the reduced glutathione levels and the enzyme glutathione-S-transferase [32, 33]:

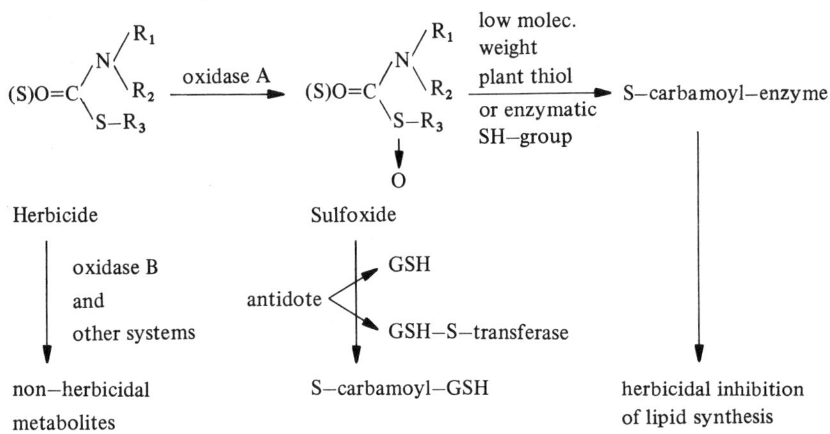

GSH = reduced glutathione

The consequence is an increased metabolism of the herbicidal sulfoxide, which is detoxified by the reaction with glutathione or other related SH-compounds.

A series of experiments has been reported indicating a requirement of light for the herbicidal activity of thiolcarbamates [34, 35]. However, the requirement of light could merely reflect that light is necessary for the conversion to the thiolcarbamate sulfoxide.

IV. Interference of Thiolcarbamate Herbicides with Photosynthetic and Respiratory Systems

Although thiolcarbamates have not been found to be efficient inhibitors of photosynthetic or respiratory systems, some reports describe inhibitory actions of these herbicides at rather high concentrations. The pI_{50}-value for electron transport inhibition in isolated chloroplasts is 3.6 for *CABAC 39* [36] and 4.4–4.6 for *di-allate* [38, 39]. *CABAC 39* and *di-allate* also affect mitochondria at a concentration of 0.1 mM [36, 39]. *Benthiocarb* has not been found to inhibit photosynthesis and respiration at the concentrations studied but affects protein synthesis [40].

V. Inhibition of Lipid Synthesis by Herbicides Other than Thiolcarbamates

A recent review summarizes the effects of herbicides on the synthesis of plant cell membrane lipids [41]. This review includes, among other things, permeability studies, model membrane studies, screening-type test systems like the betacyanine efflux from aged red beet tissue, and toxic effects of herbicides on established membranes, specifically after interference with photosynthetic reactions. Many of these aspects are dealt with in other sections of this monograph (inhibition of desaturases, p. 101; interference with membrane lipids, p. 117; relation between linoleic acid unsaturation, membrane fluidity and cold hardiness, p. 108). Here, some other structurally not related herbicides, which have been reported to inhibit fatty-acid synthesis from ^{14}C-precursors, should briefly be described.

Dinoben, *CDAA*, and *CDEC* have been observed to inhibit malonic acid–2-^{14}C incorporation into *Sesbania* hypocotyls at a concentration of 1 $mg\,l^{-1}$. *CDAA* and *CDEC* both belong to the α-chloroacetamide herbicides and may inhibit lipid synthesis indirectly in this system. *Dinoben* inhibited lipid synthesis by nearly 90% at a concentration of 1 $mg\,l^{-1}$ and was the most potent inhibitor in this test series [42]. In an in vitro system from soybean seedlings, *dinoben* inhibited the incorporation of ^{14}C-acetate and -malonate, but not that of the respective coenzyme A derivatives [43]. An inhibition of thiokinase-systems was therefore suggested as the mode of inhibitory action of *dinoben* (Table 43).

An inhibition of respiration in isolated mitochondria was only found in the 0.1–1 mM concentration range of *dinoben* [44].

Other compounds possibly interfering with lipid synthesis are *dalapon* and *TCA* (Table 43). Early data have indicated that here too, as in the case of thiolcarbamate herbicides, the formation of epicuticular waxes is affected [45, 46]. Another early observation is the partial antagonism between *dalapon* and panthothenate or β-alanine [46, 47]. Pantoic acid and β-alanine both take part in the biosynthesis of coenzyme A. However, *dalapon* did not inhibit lipid synthesis in *Sesbania* at 20 $mg\,l^{-1}$ [42], nor did *TCA* inhibit the incorporation of ^{14}C-acetate in pea seedlings [15]. Also, *TCA* did not significantly decrease the epicuticular wax formation on pea leaves [4]. Since, as is evident from the foregoing, recent studies have not confirmed the earlier data, further experiments on *dalapon* and *TCA* are needed before these compounds are really confirmed to act as inhibitors of some step of lipid metabolism.

References for G

1. Cartwright, P.M.: General growth responses of plants. In: Herbicides-physiology, biochemistry, ecology. Andus, L.J. (ed.), p. 55. London, New York, San Francisco: Academic Press 1976
2. Shibayama, H., Worley, J.F.: Weed Sci. *24*, 276 (1976)
3. Banting, J.D.: Weed Sci. *18*, 80 (1970)
4. Still, G.G., Davis, D.G., Zander, G.L.: Plant Physiol. *46*, 307 (1970)
5. Wilkinson, R.E., Hardcastle, W.S.: Weed Sci. *18*, 125 (1970)
6. Davis, D.G., Dubabek, K.E.: Weed Sci. *21*, 16 (1973)
7. Flore, J.A., Bukovac, M.J.: Hort. Sci. *11*, 316 (1976)

8. Wilkinson, R.E., Karunen, P.: Ann. Bot. *40*, 1043 (1976)
9. Wilkinson, R.E., Karunen, P.: Weed Res. *17*, 335 (1977)
10. Wilkinson, R.E.: Plant Physiol. *53*, 269 (1974)
11. Leavitt, J.R.C. et al.: Plant Physiol. *59*, 32 (1977)
12. Karunen, P., Wilkinson, R.E.: Physiol. Plant. *35*, 228 (1975)
13. Karunen, P., Eronen, L.: Physiol. Plant. *40*, 101 (1977)
14. Wilkinson, R.E. et al.: J. Agr. Food Chem. *27*, 900 (1979)
15. Harwood, J.L., Stumpf, P.K.: Arch. Biochem. Biophys. *142*, 281 (1971)
16. Sumida, S., Ueda, M.: Plant Cell Physiol. *14*, 781 (1973)
17. Kolattukudy, P.E., Brown, L.: Plant Physiol. *53*, 903 (1974)
18. Wilkinson, R.E., Smith, A.E.: Weed Sci. *23*, 100 (1975)
19. Wilkinson, R.E., Smith, A.E.: Phytochem. *15*, 841 (1976)
20. Bolton, P., Harwood, J.L.: Phytochem. *15*, 1507 (1976)
21. Wilkinson, R.E., Ashley, D.: Weed Sci. *27*, 270 (1979)
22. Chen, T.M., Seaman, D.E., Ashton, F.M.: Weed Sci. *16*, 28 (1968)
23. Donald, W.W., Harvey, R.G.: Weed Abstr. *27*, no. 1369 (1978)
24. Beste, C.E., Schreiber, M.M.: Weed Sci. *20*, 4 (1972)
25. Beste, C.E., Schreiber, M.M.: Weed Sci. *20*, 8 (1972)
26. Casida, J.E., Gray, R.A., Tilles, H.: Science *184*, 573 (1974)
27. Schuphan, I., Casida, J.E.: J. Agr. Food Chem. *27*, 1060 (1979)
28. Schuphan, I. et al.: Z. PflKrankh. PflSchutz, Sonderh. IX, 163 (1981)
29. Rosen, J.D. et al.: J. Agr. Food Chem. *28*, 880 (1980)
30. Schuphan, I., Rosen, J.D., Casida, J.E.: Science *205*, 1013 (1979)
31. Balasimha, D., Tewari, M.N., Ram, G.: Biochem. Physiol. Pflanzen *171*, 43 (1977)
32. Lay, M.M., Hubbell, J.P., Casida, J.E.: Science *189*, 287 (1975)
33. Leavitt, J.R.C., Penner, D.: Weed Sci. *27*, 47 (1979)
34. Wilkinson, R.E.: Pestic. Biochem. Physiol. *8*, 208 (1978)
35. Wilkinson, R.E.: Bot. Gaz. *138*, 270 (1977)
36. Wakamori, S.: Agr. Biol. Chem. *37*, 307 (1973)
37. Yocum, C.F., Ikuma, H.: Plant Physiol. Suppl. *56*, no. 253 (1975)
38. Robinson, S.J. et al.: Plant Physiol. *60*, 840 (1977)
39. Tetley, R.M., Ikuma, H., Hayashi, F.: Plant Physiol. *56*, 67 (1975)
40. Kimura, I., Ichizen, N., Matsunaka, S.: Weed Res. Japan *12*, 54 (1971)
41. Rivera, C.M., Penner, D.: Resid. Rev. *70*, 45 (1979)
42. Mann, J.D., Pu, M.: Weed Sci. *16*, 197 (1968)
43. Muslih, R.K., Linscott, D.L.: Plant Physiol. *60*, 730 (1977)
44. Foy, C.L., Penner, D.: Weeds *13*, 226 (1965)
45. Holmsen, T.W., Bidlack, H.D.: Proc. Weed Sci. Soc. Amer, no. 204 (1971)
46. Hilton, J.L., Jansen, L.L., Hull, H.M.: Ann. Rev. Plant Physiol. *14*, 353 (1965)
47. Hilton, J.L. et al.: Weeds *7*, 381 (1959)

H. Herbicidal Germination Inhibitors with Unknown Mode of Action

I. Metabolic Pathways Involved in the Germination and Growth of Seedlings as a Target for Herbicidal Action

In principle, each individual biochemical step important for the germination and growth of seedlings or for the regulation of these processes may be a target for herbicidal action. Unfortunately, our present knowledge of the physiology and biochemistry of seedling and cell growth sets definite limits to our understanding of the large number of mechanisms involved in the inhibition of germination by herbicides. We know a number of metabolic reactions and subcellular systems well enough to estimate their importance as a primary target for herbicides. However, we are certainly largely ignorant of many other systems.

An attempt is made here to list some of the most important physiological processes occurring during germination and growth of seedlings, as seen from the viewpoint of possible herbicidal actions (Tables 45, 46). The first compilation contains some in vivo systems where germination-inhibiting herbicides may be inhibitory. Since these systems are of the "supercomplex" type (Table 1 b), they may only occasionally yield information on the mode of action, but they may be helpful in connection with other test results. Inhibition in one or severel of these tests will mostly classify a "germination inhibitor", e.g. a high sensitivity of root growth in the case of *diphenamid* [1] or in the case of benzyl ethers [2]. Meristematic sensitivity is a phenomenon common to most herbicides acting on germination, e.g. to *alloxidim-Na* [3], to benzyl ethers [2], to *chlorsulfuron (DPX 4189)* [4], but also to the contact herbicide *chlorfenprop-methyl* (p. 182). Plant hormone-induced responses might provide valuable information, depending on the concentration required for inhibition. They form rather complex systems in that they comprise many individual steps like e.g. energy conservation, nucleic acid and protein synthesis, membrane integrity, etc. Whereas interference with cytokinin-dependent systems has not commonly been investigated – e.g. *perfluidone* [5] –, *IAA* induced *Avena* coleoptile elongation – e.g. for *alachlor [6], and propachlor* [7] – and GA_3 induced α-amylase formation in barley half seeds – e.g. for *alachlor* [8, 9] – are usually investigated when searching for herbicidal modes of action (Table 45). A decrease in the mitosis index may indicate microtubular interference, but may also originate from other primary blocks, as is the case for *perfluidone* [10] and *alachlor* [11].

The second compilation of metabolic pathways or processes (Table 46) still contains largely "complex systems" (Table 1 b) which, however, may contain the system primarily affected. The results obtained from these tests cannot, of course,

Table 45. Systems for the detection of germination- and growth-inhibiting activities of herbicides

System	Example	
	Herbicide	Page
Seedling root growth	Trifluralin	129
IAA-induced coleoptile elongation	Chlorfenprop-methyl	181
GA_3-induced α-amylase synthesis	Chlorpropham Alachlor	137 154
Cytokinin-controlled cell proliferation	Phenoxy-compounds	164
Mitosis index in root apices	Trifluralin	129
Meristematic activities: cell division and growth	Metolachlor	152

Table 46. Metabolic pathways and processes involved in germination and growth of seedlings and their relevance as a target for herbicide action

System/process	Example	
	Herbicide	Page
Hormonal regulations	2,4-D	159
Biosynthesis of hormones	EPTC	144
Respiratory energy conservation	Nitrophenols	115
Nucleic acid and protein synthesis	MDMP	120
Synthesis of specific proteins and their regulation	Alachlor	154
Biosynthesis of amino acids, pyrimidines, purines, etc.	Glyphosate	184
Biosynthesis of coenzymes, cytochromes, etc.	?	–
Lipid synthesis and membrane organization	EPTC Trifluralin	142 117
Microtubular synthesis and control systems	Chlorpropham	129
Ion transport and membrane permeability	Alachlor	154

definitely prove a primary target of a herbicide because of the many reactions involved. Tracer studies and hormone test systems of different types have generally been applied with most herbicides, and also with the compounds of interest here. The list can certainly not be complete.

Whereas the basic biosynthetic pathways are fairly well known in many cases, the underlying regulatory processes are still largely a "black box". It is only intended here to list some of the more important biochemical pathways which have to be considered when dealing with germination-inhibiting herbicides. Emphasis should not only be laid on the concentration dependencies of possible inhibitions, but also on the time lag observed between herbicidal addition and eventual inhibition, in order to discriminate between direct and indirect effects.

II. Germination-Inhibiting Herbicides with Unknown Primary Biochemical Mode of Action

1. α-Chloroacetamides

The more important presently used α-chloroacetamide herbicides are compiled in Table 47. Most of these compounds are α-chloroacetanilides. However, the herbicide needs not necessarily contain an aromatic ring as evidenced by *CDAA*. The α-chlorine atom, on the other side, must be present to induce herbicidal activity [12]. For the α-chloroacetanilide series, it could be demonstrated that they occur in two isomeric/rotameric forms, due to steric hindrance of the o-substituent on the ring [12a]:

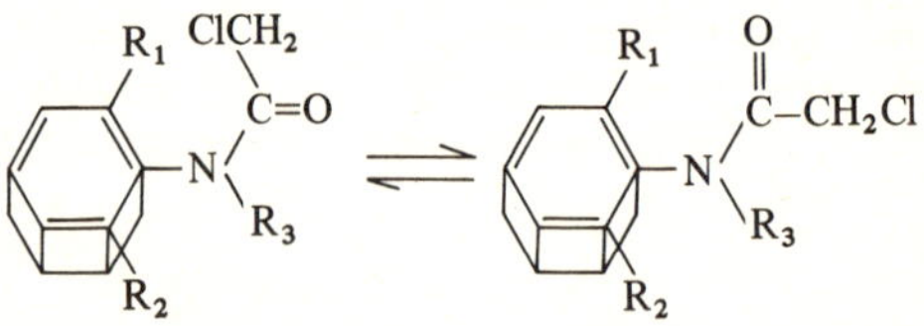

A B

i.e. the >N–$COCH_2Cl$ linkage has partial double-bond character. At equilibrium, the ratio of A: B will be around 10 for most compounds. However, equilibrium appears often to be reached only slowly so that the two rotamers can be isolated individually in some cases [12]. In these examples it could also be demonstrated, that B is the herbicidally active form. Since this is also the rotamer with a high alkylation activity, herbicidal activity of α-chloroacetamides has been suggested to originate from an alkylation step. This would mean a firm binding of the herbicide to the site of action.

Attempts to detect the mode of herbicidal action for these compounds will be summarized in "growth analysis" and "effects on cellular metabolism".

a. Analysis of the Growth of Plants and Tissues Treated with α-Chloroacetamides

Inhibition of growth after application of α-chloroacetamides always occurs only after a time-lag of several hours. The period of this time-lag lies between 3–4 h in the *Avena* coleoptile straight growth test [6] and 48 h in pea and oat roots [11], depending on the herbicidal concentration and on the type of tissue. Of course, the time-lag decreases with increasing herbicide concentration. At very high herbicidal concentration the risk of inducing growth inhibition from several primary inhibitions will certainly increase. Balke [13, 14] and Watson et al. [15] found that at 0.1–1 mM *alachlor*, membrane permeability was affected in oat roots and oat mesophyll protoplasts, leading to a decreased ion absorption and membrane leakage,

Table 47. Important α-chloroacetamide herbicides

Common name	Structure	Water sol. (mg l^{-1})
Acetochlor	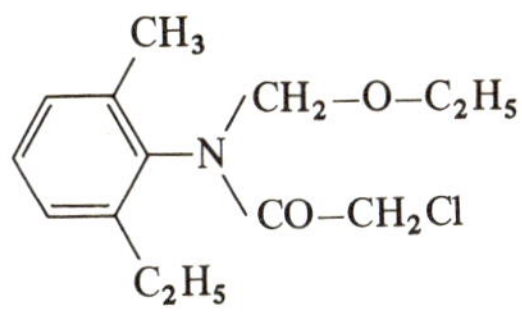	–
Alachlor	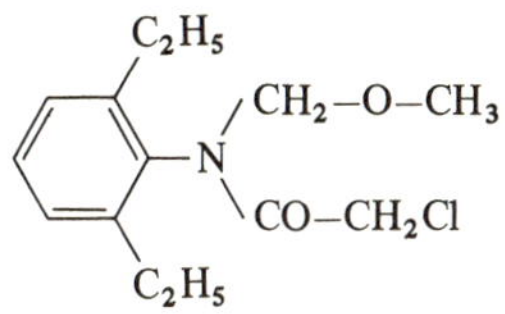	240
CDAA = allidochlor	$CH_2=CH-CH_2-N(CH_2-CH=CH_2)(CO-CH_2Cl)$	(19 g l^{-1})
Butachlor	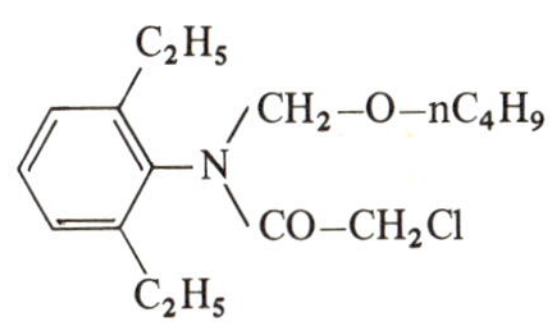	20
Metolachlor	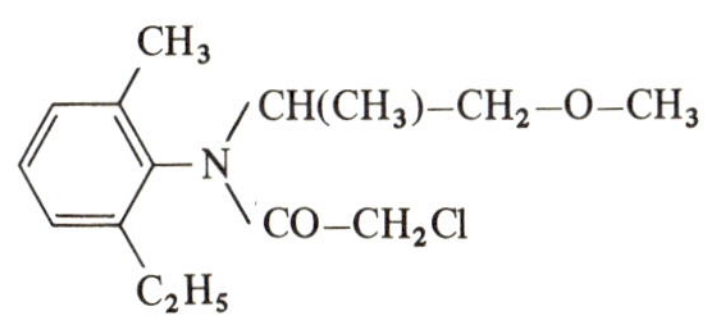	530
Propachlor	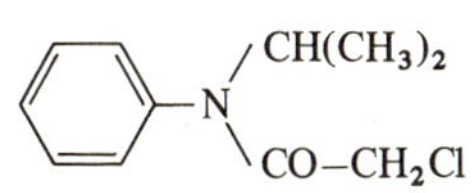	700
Prynachlor	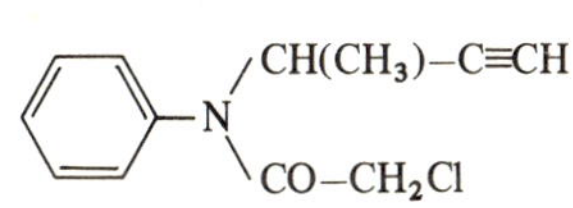	–

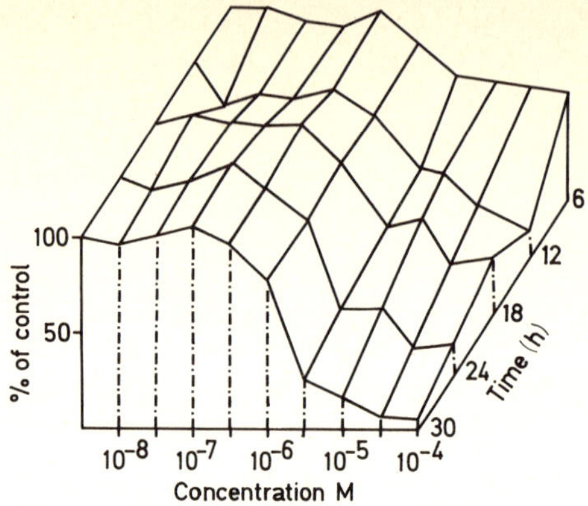

Fig. 40. Growth analysis of germinating oats in the presence of increasing concentrations of *metolachlor* [11]. Reproduced with permission of the Weed Science Society of America

respectively. These effects have yet to be correlated with the growth inhibitions occurring at lower concentrations.

Effects on metabolite uptake and transport in oat seedling roots have been correlated with growth inhibitions after treatment with *alachlor* [16]. An early effect on membrane transport (15 min) and a later inhibition of protein synthesis (2 h) have been distinguished. Root elongation, cell enlargement and mitotic activity are then also inhibited. Inhibition of root elongation and cell division have also been reported for *propachlor* [7, 17].

An interesting finding is the antagonism between GA_3-treatment and *alachlor* treatment in oats and corn [18, 19]. In both cases the tissue is less responsive to *alachlor* when pretreated with GA_3; a simultaneous treatment in oats showed no GA_3-influence. In corn, the time-lag of *alachlor* inhibition decreases from 21 h in the absence to 9 h in the presence of GA_3, and the GA_3-induced growth is inhibited by *alachlor*. Since GA_3 rapidly stimulates cell division and growth, the faster action of *alachlor* in GA_3-treated tissue again indicates an action of the herbicide on one or both of these processes. Light was also found to influence the activity of *alachlor* in oat seedlings under certain conditions [20].

From the growth data discussed so far, inhibition of cell division and cell enlargement may be concluded as the effect first visible at the cellular level (Fig. 40). At excessive concentrations of *alachlor* and *butachlor*, cytogenetic irregularities have been found [21]. The physiological meaning of these data is, however, questionable.

Another indication of the action of *alachlor* on cell division comes from experiments with the unicellular green alga *Chlamydomonas reinhardii* (Fig. 38) [22]. The growth inhibition found in this alga could be attributed to an inhibition of cell division in synchroneously growing cultures (Table 48). The growth of cells during the environmental light phase is not inhibited. However, cell division is blocked when the herbicide has been present for several hours before the onset of the dark phase. Herbicidal addition at the beginning of the dark phase does not block division but only the separation of the daughter cells which grow to big clusters during

Table 48. Action of 10 mg l^{-1} *alachlor* on *Chlamydomonas reinhardii* growth and multiplication. Alachlor was added to the culture solutions at different time intervals and removed by centrifugation and resuspension of the cells. Inoculation was performed with an algal cell volume of 90 μm^3. Light phases were 0–10, 24–34, and 48–58 h. C = control, A = *alachlor* treated [22].

Time interval of the presence of alachlor h		Average particle volume (μm^3)						Increase of particle number during time interval in %		
		8	24	32	48	56	72	0–30	30–54	54–70
–	C	181	84	156	92	183	115	268	184	210
0–72	A	200	440[a]	230	220	–	–	10	30	0
9–72	A	150	200[b]	580[b]	400[b]	–	–	60	0	0
0– 9	A	160	230[a]	500[a]	100	200	100	30	690	220
24–72	A	210	80	150	310[a]	170	150	310	20	40
0–24	A	190	440[a]	270	100	200	100	0	70	250

[a] Single cells without flagella
[b] Aggregates of 2–5 cells

the next light phase. This finding clearly indicates that *alachlor* (a) does not inhibit cell growth and that (b) cell division is inhibited only indirectly. Apparently, some preparatory step or process required for cell division is blocked. This reminds of the situation in higher plants where the inhibition is only complete when the herbicide is present from the very beginning of germination.

Chlamydomonas cells treated with *alachlor* from the beginning of the light phase prepare for cell division by shedding their flagella and swelling at the beginning of the dark phase; however, actual division does not take place. Another interesting result of the experiments recorded in Table 48 is the finding that the herbicide can be removed by simple washing 9 h after its addition to the culture solution. The immediate division during the first dark phase than does not take place. However, the cells grow on during the second light phase and divide into 7–8 daughter cells during the second dark phase, thereby making up for the missing cell divisions of the first dark phase/growth cycle.

It may be concluded from the growth data discussed in this chapter that α-chloroacetamides are indirect inhibitors of cell division in higher plants due to blocking of some regulatory or biosynthetic step needed for the normal cell division to occur. We have no idea so far what this step might be. The evidence of inhibitory actions on plant metabolism will be discussed in the next chapter.

b. Effects of α-Chloroacetamides on Cellular Metabolism

Protein synthesis. By far the largest amount of evidence has been accumulated on the inhibition of protein synthesis by α-chloroacetamides. *Alachlor*, *metolachlor*, *propachlor*, *prynachlor*, and *CDAA* have been shown to inhibit protein synthesis and polyribosome formation [23–26]. However, as for the growth inhibition described above, a lag phase was observed between herbicide addition and the onset of the inhibition of polyribosome formation. Also, in vitro protein synthesis was not inhibited by 0.1 mM *alachlor* [25]. All these findings therefore mean that, although protein synthesis appears to be one of the cellular processes affected early after herbicidal application, it is not the directly sensitive biochemical pathway. It should also be considered that possibly only the synthesis of a specific group of proteins needs to be inhibited in the course of herbicidal action.

Protein synthesis is also involved in the GA_3-induced α-amylase formation in barley seeds, which is sensitive to several α-chloroacetamides at concentrations above 10 μM [8, 9, 26].

Ion transport and membrane function. As discussed before, concentrations of *alachlor* higher than 0.1 mM affect the membrane function, measured as ion uptake or solute leakage, in a more direct way. These effects might be responsible for the sensitivity of the *Avena* coleoptile elongation. It is, however, not at all clear whether these inhibitions are related to the herbicidal activity. The coleoptile elongation does not include cell divisions, whereas, in contrast, α-chloroacetanilides at μM concentrations appear to affect primarily cell division.

Lipid synthesis. Mann et al. [27] observed a strong inhibition of ^{14}C-malonate incorporation into the lipid fraction by 1 $mg\,ml^{-1}$ CDAA. In a more recent study, Truelove et al. [28] reported the specific inhibition of ^{14}C-choline-incorporation into phosphatidyl choline by *metolachlor* and *alachlor* in cotton root tips. Also, lipid synthesis in sorghum coleoptiles [54] and ^{14}C-mevalonate metabolism in microsomes from the same tissue [55] were inhibited by *metolachlor*. These inhibitions might well relate to the inhibitory effects on membrane function and protein synthesis and should be studied in more detail.

Alkylation. As mentioned above, the molecular mode of herbicidal action has been suggested to be an alkylation step [12]. Jaworski [26] applied this idea to the inhibition of protein synthesis and suggested an alkylation of specific m-RNAs. The herbicidal action should then be irreversible. However, the herbicide may be washed out of inhibited algal cells which will then resume growth [22]. On the other hand, the metabolic breakdown of α-chloroacetanilides includes an early loss of the α-chlorine. This breakdown, however, might be due to the reaction with SH-compounds, as described below [30], and may or may not be connected with the biochemical mode of action of these herbicides.

Other metabolic pathways. Little information is available on the influence of α-chloroacetamides on other metabolic pathways. Photosynthesis is not sensitive to these herbicides [23]. Respiration has been found to be inhibited by *CDAA* in vivo [26]. This inhibition should, however, occur indirectly. Kleudgen [29] has analyzed the chloroplast pigments in *propachlor*-treated barley seedlings. He came to the conclusion that *propachlor* antagonized the effects of exogenous *IAA* and induced a state of senescence in the chloroplast.

SH-reactivity. α-chloroacetanilides react spontaneously with SH-compounds via their α-chlorine in a nucleophilic displacement reaction. The SH-compound most readily available in the cell for this type of reaction appears to be reduced glutathione (GSH). This may be concluded from the fact, that the herbicide antidote R 25,788 causes an increase of the GSH-level in corn [31]. *Metolachlor*, *alachlor*, and *propachlor* were all demonstrated to form the respective GSH-conjugate in vitro [30, 32].

Although α-chloroacetamides readily react with SH-compounds like glutathione, they cannot be classified as a general SH-reagent. When several SH-enzymes were tested in the presence of 0.3 mM *alachlor*, no inhibition was observed [33].

2. Other Germination-Inhibiting Herbicides

The structures of a number of pre emergence herbicides, for which only very little information on their mode of action is available, are given in Table 49. As pre emergence applied herbicides they will conceivably primarily affect meristematic tissues.

Alloxidim-Na. This is a herbicide acting primarily on gramineous plants. The meristematic tissue appears to be most sensitive [34, 34a]. This herbicide does not inhibit protein synthesis [35]. Extensive studies on the uptake, translocation and degradation of *alloxidim-Na* have been published [36–38].

Diphenamid. *Diphenamid* is only acting on roots where cell division appears to be blocked. This specific sensitivity reminds of the microtubule-inhibiting herbicides; however, in the presence of *diphenamid* the number of meristematic cells showing mitosis did not increase [39]. In a study with corn roots Yaklich et al. [40]

Table 49. Germination-inhibiting herbicides with unknown primary mode of action (α-chloroacetamides excluded)

Common name	Structure	Water sol. (mg l^{-1})
Alloxidim-Na = carbodimedon	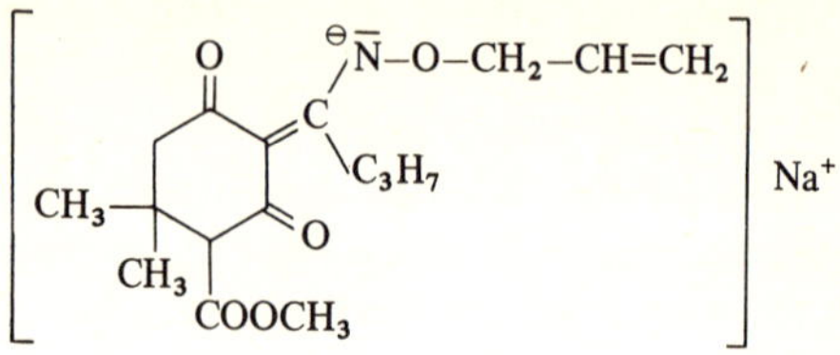	(2 kg l^{-1})
Diphenamid	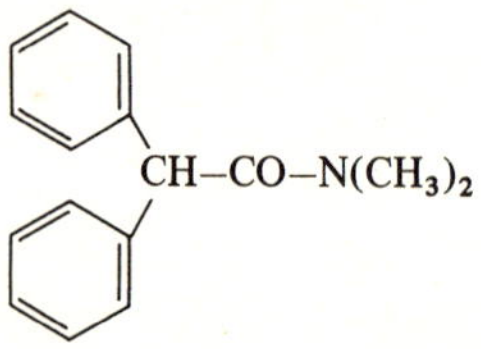	260
Chlorsulfuron	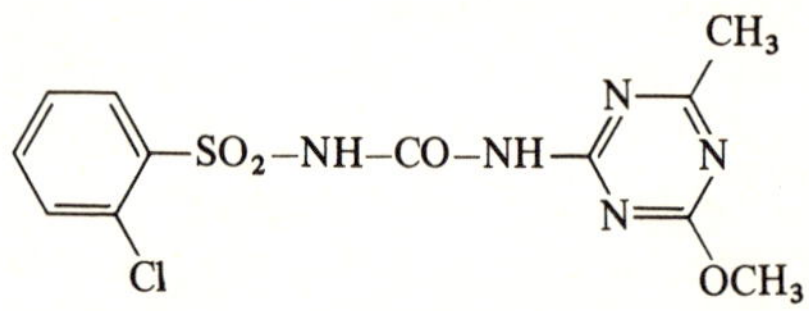	–
Perfluidone	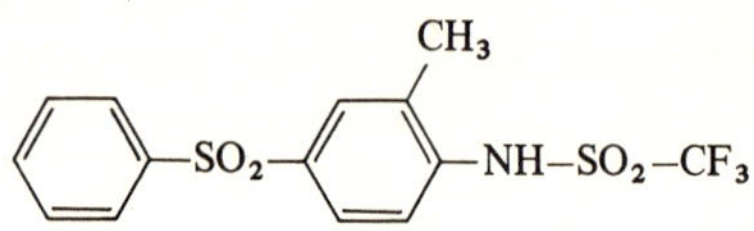	60
Benzyl ethers	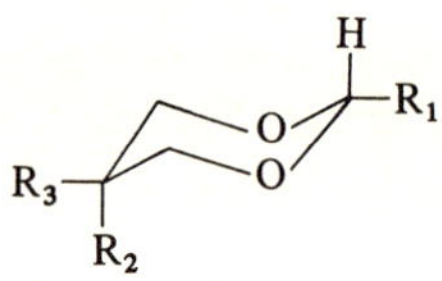	–
Glycerol ethers	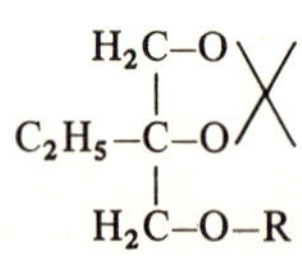	–

tried to locate the site of action by subcellular distribution studies. They were, however, not successful. The metabolism has been studied in soybean cell suspension cultures [41].

Chlorsulfuron (DPX 4189). This very interesting new herbicide is active at extremely low concentrations (10–60 g a.i./ha). Only some preliminary information can be reported on its mode of action which will certainly be studied in more detail in the near future. The compound *chlorsulfuron* rapidly inhibits the incorporation of thymidine into the DNA of corn seedling roots, i.e. by 60–70% at a concentration of 2.8×10^{-6} M after 2 h [42]. With the same type of incubation, cell elongation, protein and RNA synthesis, photosynthesis, and respiration were not inhibited. When the percent distribution among different nuclear division phases was checked, no deviation from controls could be detected. The preliminary conclusion therefore is, that *chlorsulfuron* acts as a very specific inhibitor of cell division in higher plants.

With isolated *Phaseolus vulgaris* leaf cells, de Villiers et al. [43] found inhibition of photosynthesis, respiration and of the syntheses of protein, RNA and lipids at 0.5 mM *chlorsulfuron*. This concentration, however, is too high to draw any conclusions on the herbicidal mode of action.

Perfluidone. This herbicide exhibits two different activities which are not necessarily correlated. An influence on in vivo IAA-oxidase activity [44] may be related to the hormone type of activity: release of bud dormancy and apical dominance with simultaneous growth inhibition and increased tuber formation in *Cyperus rotundus* have been reported [45].

The primary mode of the second type of action, probably the herbicidal, is equally obscure. Photosynthesis was not inhibited by 10 μM *perfluidone* in peanut protoplasts [46]. At higher concentrations in vitro *perfluidone* acts as an uncoupler of photosynthetic electron transport [47]. A high sensitivity to *perfluidone* was observed for mitosis (1–10 μM) [48] and for the synthesis of lipids (0.1 mM) [49].

Lamoureux et al. [50] have studied the uptake, transport and metabolism of *perfluidone* in peanuts.

Benzyl ethers and glycerol ethers. This new group of related pre emergence herbicides needs the meristematic contact in order to become herbicidally active [51]. Root growth is highly sensitive, but other meristems appear to be affected, too. Whereas photosynthesis, respiration, and also RNA-synthesis were unaffected in susceptible corn plants, DNA- and protein synthesis proved to be very sensitive [52]. As a consequence, cell division is strongly inhibited. Burrows [53] also observed a high sensitivity of DNA-synthesis to herbicides of this group. However, in his experiments, RNA-synthesis was more sensitive than protein synthesis. He suggested a herbicidal effect on the integrity of the chromatin complex.

References for H

1. Deli, J., Warren, G.F.: Weed Sci. *18*, 692 (1970)
2. Burrows, W.J.: Pestic. Biochem. Physiol. *8*, 137 (1978)
3. Iwataki, I., Hirono, Y.: The chemical structure and herbicidal activity of alloxidim-Na and related compounds. In: Proc. fifth Int. Congr. Pestic. Chem. (IUPAC), no. II-12, Zürich 1978
4. Ray, T.B.: Studies on the mode of action of DPX-4189, Proc. Brit. Crop Prot. Conf.-Weeds, p. 7 (1980)
5. Bendixen, L.E.: Weed Sci. *23*, 445 (1975)
6. Bardzik, J.M.: Weed Abstr. *24*, no. 2778 (1975)
7. Dhillon, N.S., LaMar Anderson, J.: Weed Res. *12*, 182 (1972)
8. Rao, V.S., Duke, W.B.: Weed Sci. *24*, 616 (1976)
9. Devlin, R.M., Cunningham, R.P.: Weed Res. *10*, 316 (1970)
10. Davis, D.G.: Weed Sci. *23*, 81 (1975)
11. Deal, L.M., Hess, F.D.: Weed Sci. *28*, 168 (1980)
12. Hamm, P.C.: Weed Sci. *22*, 541 (1974)

12a. Hamm, P.C.: Some unique biological activity-structure relationships of the acylated anilides of the alachlor type. In: Herbicides, fungicides, formulation chemistry, Proc. 2. Int. Congr. Pestic. Chem. (IUPAC), Vol. V. Tahori, A.C. (ed.), p. 41. New York, London, Paris: Gordon and Breach 1972
13. Balke, N.E.: Abstr. Weed Sci. Soc. Amer. Meet., no. 198 (1979)
14. Balke, N.E.: Abstr. Weed Sci. Soc. Amer. Meet., no. 183 (1980)
15. Watson, M.C., Bartels, P.G., Hamilton, K.C.: Weed Sci. *28*, 122 (1980)
16. Marsch, H.V., Bates, J., Downs, S.: Plant Physiol. Suppl. *57*, no. 319 (1976)
17. Duke, W.B. et al.: Weed Sci. *23*, 142 (1975)
18. Chang, T.C., Marsh, H.V., Jennings, P.H.: Pestic. Biochem. Physiol. *5*, 323 (1975)
19. Narsaiah, D.B., Harvey, R.G.: Weed Sci. *25*, 197 (1977)
20. Chang, T.C., Marsh, H.V., Jennings, P.H.: Plant Physiol. Suppl. *49*, no. 253 (1972)
21. Das, K. et al.: Weed Abstr. *29*, 98 (1980)
22. Fedtke, C.: Mode of action information for herbicides from using *Chlamydomonas reinhardii* and a Coulter Counter. In: ACS Symposium Series, in press, 1981
23. Chandler, J.M., Croy, L.I., Santelman, P.W.: J. Agr. Food Chem. *20*, 661 (1972)
24. Rao, V.S., Duke, W.B.: Weed Abstr. *27*, no. 4216 (1978)
25. Deal, L.M. et al.: Weed Sci. *28*, 334 (1980)
26. Jaworski, E.G.: J. Agr. Food Chem. *17*, 165 (1969)
27. Mann, J.D., Pu, M.: Weed Sci. *16*, 197 (1968)
28. Truelove, B. et al.: Abstr. Weed Sci. Soc. Amer. Meet., no. 210 (1979)
29. Klendgen, H.K.: Weed Res. *20*, 41 (1980)
30. Frear, D.S. et al.: Adv. in Agron. *24*, 327 (1972)
31. Lay, M.-M., Casida, I.E.: Pestic. Biochem. Physiol. *6*, 442 (1976)
32. Leavitt, J.R.C., Penner, D.: J. Agr. Food Chem. *27*, 533 (1979)
33. Marsh, H.V., Bates, J., Trudeau, P.: Abstr. Weed Sci. Soc. Amer. Meet., p. 67 (1975)
34. Iwataki, I., Hirono, Y.: The chemical structure and herbicidal activity of alloxidim-Na and related compounds. In: Abstr. Vol. fifth Int. Union for Pure and Appl. Chem. (IUPAC) Congr., no. II-12, Zürich 1978
34a. Ishikawa, H. et al.: J. Pestic. Sci. *5*, 547 (1980)
35. Bugg, M.W., Hayes, R.M.: Abstr. Vol. Southern Weed Sci. Soc. Amer. Meet., 245 (1978)
36. Soeda, Y. et al.: J. Pestic. Sci. *4*, 121 (1979)
37. Hashimoto, Y., Ishihara K., Soeda, Y.: J. Pestic. Sci. *4*, 299 (1979)
38. Hashimoto, Y., Ishihara, K., Soeda, Y.: J. Pestic. Sci. *4*, 375 (1979)
39. Deli, J., Warren, G.F.: Weed Sci. *18*, 692 (1970)
40. Yaklich, R.W., Scott, E.G.: Physiol. Plant. *28*, 447 (1973)
41. Davis, D.G. et al.: Physiol. Plant. *44*, 87 (1978)
42. Ray, T.B.: Studies on the mode of action of DPX-4189. In: Proc. Brit. Crop Prot. Conf.-Weeds, p. 7 (1980)
43. de Villiers, O.T., Vandenplas, M.L., Koch H.M.: The effect of DPX-4189 on biochemical processes in isolated leaf cells and chloroplasts. In: Proc. Brit. Crop Prot. Conf.-Weeds, p. 237 (1980)
44. Tafuri, F. et al.: J. of the Sci. of Food and Agr. *28*, 180 (1977)
45. Bendixen, L.E.: Weed Sci. *23*, 445 (1975)
46. Davis, D.G., Shimabukuro, R.H.: Weed Abstr. *25*, 148 (1976)
47. Alsop, W.R., Moreland, D.E.: Pestic. Biochem. Physiol. *5*, 163 (1975)
48. Davis, D.G.: Weed Sci. *23*, 81 (1975)
49. St. John, J.B., Hilton, J.L.: Weed Sci. *21*, 477 (1973)
50. Lamoureux, G.L., Stafford, L.E.: J. Agr. Food Chem. *25*, 512 (1977)
51, Jeffcoat, B.: Pestic. Sci. *8*, 43 (1977)
52. Templeton: Dissertation, Virginia Univ. (1976)
53. Burrows, W.J: Pestic. Biochem. Physiol. *8*, 137 (1978)
54. Ebert, E.: Pestic Biochem. Physiol. *13*, 227 (1980)
55. Wilkinson, R.E.: Abstr. Weed. Sci. Soc. Amer. Meet., no. 235. (1981)

I. Herbicides with Auxin Activity

I. Tests for Auxin Activity

Here we do not want to discuss the physiology of auxins, nor the mode of action of auxin hormones, although both aspects are nearly identical with the question for the molecular mode of action of auxin herbicides. Many reviews are mainly concerned with auxins, necessarily including auxin herbicides [1–6]. In a recent critical review Trewavas [142] suggested a new theory where not the concentration of the endogenous auxin but the sensitivity of the tissue (receptor availability) is the controlling factor. This idea would equally apply to other plant growth substances, and the term plant hormone – coined from the anticipated analogy with animal hormones – would then be inadequate. This chapter intends to give a comprehensive overview of this class of herbicides and then to concentrate on some relevant aspects.

One basic question to be asked would of course be the following: how can a herbicide be recognized as an artificial auxin? Auxins, as other plant hormones too, are stimulatory in a number of test systems, and these stimulations can serve as a measure of hormone activity. However, stimulation is a matter of concentration, and plant hormone stimulations therefore display optimum concentration curves. An example of this type of response is given in Fig. 41 which also illustrates the second peculiarity of plant hormone growth responses: the different sensitivity of different tissues. With these two peculiarities in mind – optimum concentration curves

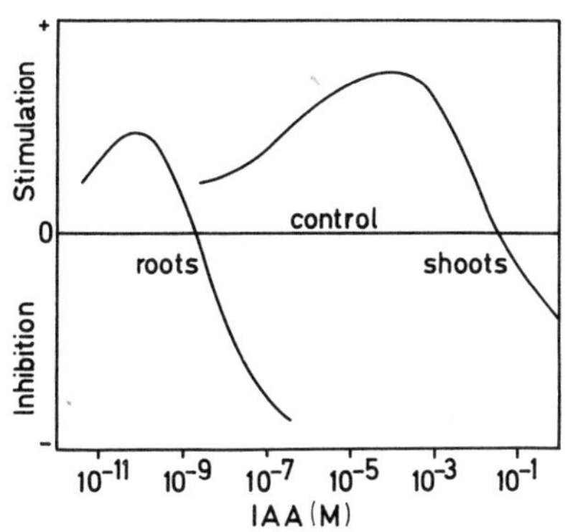

Fig. 41. Increase in length in roots and shoots influenced by externally applied IAA. Control: endogenous IAA only

Table 50. Herbicidal phenoxy acids

Common name	Structure	Water sol. (mg l^{-1})
2,4-D	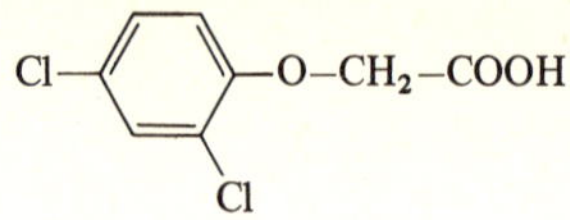	620
2,4-DB	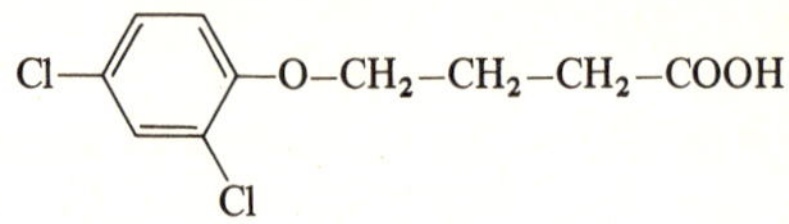	46
2,4-DP = dichloprop	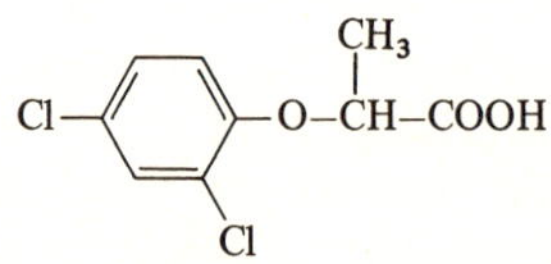	3,500
MCPA	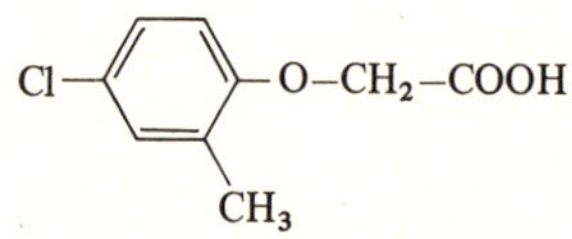	825
2,4,5-T	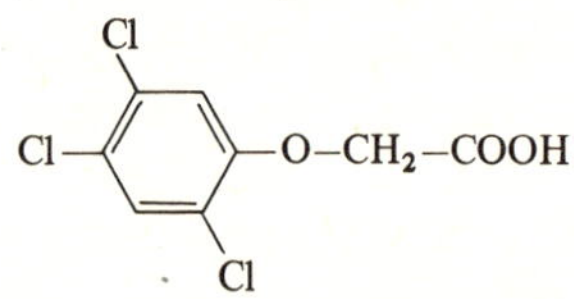	238

and tissue specificity of the response – hormone-type growth responses to chemicals can now be studied in more detail.

Oat coleoptile elongation: the oat coleoptile elongates in response to low concentrations of *IAA*. Since this elongation is only based on cell elongation – no cell divisions, no interplay with other plant hormones –, the test is ideally suited for the demonstration of auxin activity. In effect, it has been through a screen of various chemicals with this test system that the most common artificial auxin *2,4-D* (Table 50) has originally been found [5]. Using the oat coleoptile system, four different aspects of auxin action can be measured which can all be checked with the test chemical in question: (a) proton (H^+) excretion leading to the activation of cell wall hydrolases [139], specifically β-1,3-glucanase [7], which reduce the strength of the wall and allow the cell turgor to push for the elongation of the cell [8, 9], (b) the coleoptile elongation proper, measurable as an elongation or as a curvature when the auxin was applied only on one side of the tip, (c) unidirectional basipetal transport, and (d) conjugation of the artificial auxin with aspartate and other amino acids. The latter process involves detoxification of the artificial auxin when added in excess amounts [5].

Bean hypocotyl elongation: the elongation of etiolated hypocotyls of mung and soy bean also responds to the presence of auxins. Stimulations or inhibitions may be found, depending on concentration. However, contradictory reports exist as to whether H^+-excretion participates in the control of the hypocotyl elongation or not [10–12]. The mung bean hypocotyl system has e.g. been employed by Vendrig [13] in his studies on *orthonil.*

Cell culture systems: plant cell cultures may be grown as suspension cultures, callus cultures or organ cultures. In most cases, plant hormones of the auxin and/or of the cytokinin type have to be included in the culture medium for the control of cell growth and multiplication. Specifically, the ratio of the concentrations of auxins and cytokinins present in the culture medium profoundly influences the type of growth. Calluses, roots or shoots may be obtained simply by varying the above ratio.

It is of interest that different artificial auxins, although substituting for the natural auxin in the cell culture system, may produce different morphological effects, as has e.g. been shown for *benazolin* [14]. In general, however, auxin herbicides fully replace *IAA* in cell culture systems, as has e.g. been demonstrated for *2,4-D* [15], *chloramben* [16], *dicamba*, and *picloram* [17].

Other auxin indicator systems: many more test systems employing tissue pieces (e.g. the pea split stem test), seeds, geotropic stem/root curvature [18], or other specific plant parts may serve as an indicator of auxin activity [19]. However, most of these systems did not gain importance in studies concerned with the mode of action of auxin herbicides. Also, the application of several of the possible systems is problematic since they are not specific to auxins.

II. Structural Requirements for Auxin Activity

Herbicides exhibiting auxin activity are classified into three subgroups: (a) "phenoxy acids" (Table 50), (b) "auxin-active compounds with an acetic acid moiety directly attached to an aromatic system" (Table 51), and (c) auxin herbicides "derived from benzoic acid and related structures" (Table 52). Whereas groups (a) and (c) contain highly active herbicides, the compounds in group (b) are more of the growth-regulator type and also comprise the natural auxin IAA. The history of these compounds also differs to some extent: whereas the phenoxy acids have originally been found in screening programs aimed at growth regulators, most of the compounds compiled in Table 52 come from routine herbicide screening. Thus, for group (a) compounds the herbicidal activity was detected after the auxin activity whereas for the group (c) compounds the reverse is true.

The question of whether an artificial auxin acts as a growth regulator or as a herbicide appears largely to be a matter of potency, stability in the tissue and transport characteristics. Although these questions are discussed more thoroughly later (p. 170), it should be mentioned here that the metabolic stability of the phenoxy acids appears to be the main reason for their herbicidal activity.

An obvious characteristic of all auxin-active structures is the free carboxy group. Compounds like *orthonil* (Table 51) are metabolized to the active free acid [20]. Some auxin-active chemicals without a free carboxy group do exist; how-

Table 51. Auxin-active compounds with a carboxymethyl group directly attached to an aromatic system

Common name	Structure
Benazolin	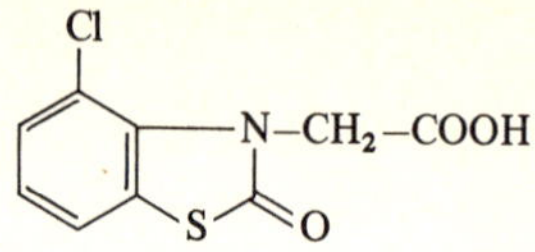
Fenac	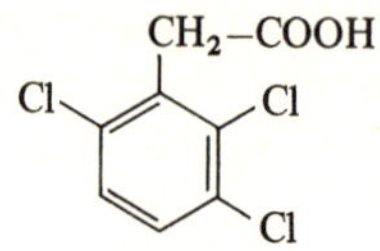
IAA	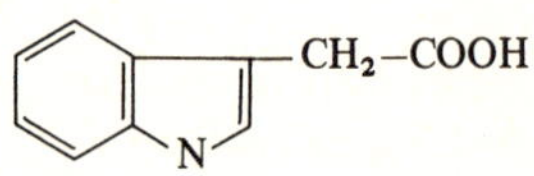
NAA	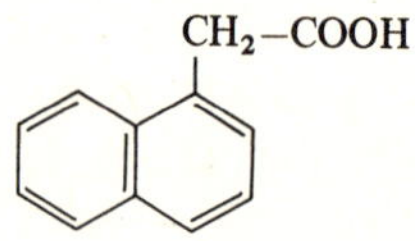
Orthonil	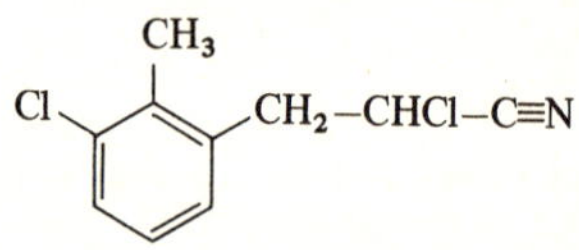

ever, a similar negatively charged substituent may also be discerned in these molecules [5]. It is therefore generally agreed that a negatively charged group is essential for an auxin-active structure. However, the minimum structural requirements for the remainder of the molecule are somewhat less obvious. In Fig. 42 two currently used models of the structural requirements for an active auxin and of the corresponding binding site on the receptor protein are illustrated. The suggestion by Thimann [5] and Farrimond [21] envisages a flat lipophilic group/ring system carrying a positive charge and a negative substituent – the carboxy group – at 5.5 Å distance from the positive charge, emerging from the plane of the ring or situated in a "particular orientation with respect to the ring" (Fig. 42a). Another view of the auxin-binding site interference holds that a positive charge is not necessary (Fig. 42b) [22a]. Instead, π-bond interaction or a charge-transfer complex might hold the ring structure at the binding site. The binding is, of course, freely reversible. Since some unsaturation in the ring and lipophilic substituents are required for maximum activity, lipophilic interference should occur during binding. However, since Katekar envisages a planar and electrophilic binding site for the ring system, both lipophilic and electrogenic components might contribute to the bind-

Table 52. Herbicides with auxin activity derived from benzoic acid and related structures

Common name	Structure	Water sol. (mg l^{-1})
Chloramben	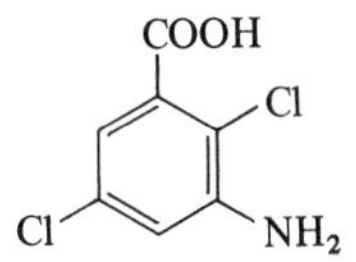	700
Dicamba	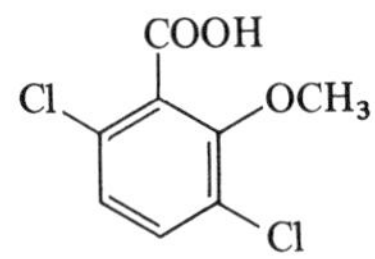	4,500
Picloram	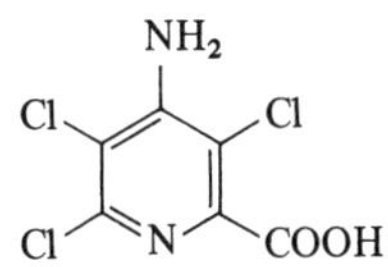	430
2,3,6-TBA	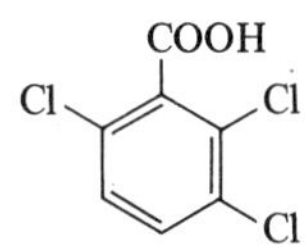	8,400
Tricamba	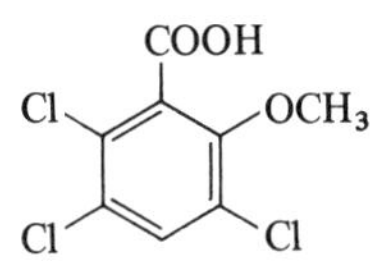	–

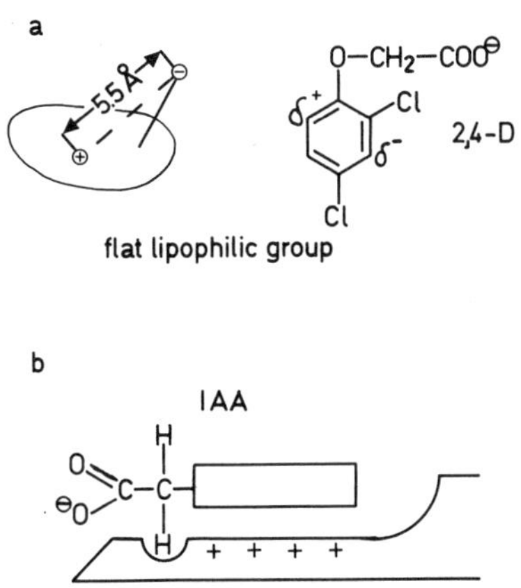

Fig. 42. Minimum structure of auxin-active compounds according to Thimann [5] **a** and side view of *IAA*-binding site interference according to Katekar [22] **b**

ing of the ring structure. As an example, the fractional charge distribution is given for *2,4-D* (Fig. 42a) to illustrate the possibility of lipophilic and electrogenic interactions with the same molecule.

There has also been put foreward a "conformational change theory for auxin structure-activity relationships" [23]. This theory envisages a planar "recognition conformation" which moves in concert with the receptor to a "modulation conformation".

III. Auxin Interaction with Other Plant Hormones

Auxin is synthesized in the meristems and leaves of plants and transported symplastically in a basipetal manner. On its way downstream through the plant the concentration rapidly decreases through oxidative and peroxidative destruction and through complex formation (see below). The remaining free *IAA* may then control growth, morphogenesis and metabolism in any place by way of its own concentration or by interaction with other plant hormones. Two interactions with other plant hormones are particularly important and will therefore be discussed here. Some consideration will also be given to the possible destruction by oxidases and peroxidases.

1. Interaction with Cytokinins

Cytokinins induce cell divisions in tissue explants. However, this effect is modulated by auxin, and, most importantly, it is the concentration ratio between cytokinins and auxin that actually controls morphogenesis [24, 25]. Although morphogenesis may be modulated by the presence of amino acids, phosphate, etc., the basic controls are as follows: induction of bud and shoot formation in the presence of high cytokinin and low auxin concentrations, callus growth without morphogenesis at intermediate concentrations of both hormones, and root formation in the presence of high auxin and low cytokinin concentrations. When adapted to the intact plant these results mean that the *IAA* moving downward from the apical meristem suppresses axillary bud growth, i.e. as long as the upper meristems are functional, branching is inhibited. The basis for this interaction might be an influence of kinetin on auxin metabolism [140].

2. Interaction with Ethylene

The gaseous plant hormone ethylene is continuously produced by most plant tissues [26]. When an auxin herbicide is applied, this production is greatly increased, as shown e.g. for *2,4-D* and *IAA* [1], *dicamba* [27], *picloram* [28], and *NAA* [29]. Since some effects of auxins and of ethylene are similar, it was originally suggested that auxins might act via the induction of excessive ethylene production. However, this is not true. Firstly, the effects on nucleic acid metabolism are not mediated by ethylene [30]. Moreover, auxins and ethylene are antagonists rather than synergists. Since the synthesis of ethylene appears, at least in part, to be controlled by the auxin concentration, the system would be self-regulating.

One important example is the abscission of leaves and fruits. In the presence of externally applied excess amounts of auxins and the artificial production of high concentrations of ethylene, fruit ripening and abscission are promoted [31]. The surface of the treated fruit, however, remains fresh and undergoes colour changes in response to the juvenating effect of the auxin. The natural abscission, on the contrary, is initiated when the endogenous *IAA* synthesis by the aged leaf fades and endogenous ethylene is then produced in the abscission zone. Similar to the sprayed fruit, the punctual treatment of a leaf with a drop of *2,4-D* produces a green treated zone and a yellow surrounding ring. The explanation again is that *2,4-D*, at the site of application, keeps the tissue juvenile whereas the mobile ethylene produced in response to the *2,4-D*-treatment diffuses out and softens and ages the surrounding tissue [1]. In summary, auxin and ethylene are in many respects antagonistic, e.g. in growth, abscission and ripening. Ethylene might also be responsible for many of the sublethal effects observed after *2,4-D* treatment (p. 169).

As already stated above, the response of tissues to auxin treatment is primarily a matter of physiological status. This also applies to the ethylene response and the correlated effects which may not always be found with every type of tissue.

3. Auxins and Oxidases/Peroxidases

As mentioned before, *IAA* is rapidly metabolized, and oxidases and peroxidases are some of the candidate enzymes for this metabolic breakdown. Unfortunately, these enzymes occur in higher plant cells as a complement of many isoenzymes whose spectra differ with the tissue [32], with the age of the cells [33, 34] and also with the hormone complement present [34, 35]; they are therefore difficult to evaluate.

Let us first consider the types of reaction under discussion. The oxidase reaction introduces one oxygen atom into an acceptor molecule AH_2 (e.g. auxin) which is converted to HAOH:

$$\text{oxidase: } AH_2 + O_2 + DH_2 \rightarrow HAOH + D + H_2O;$$

the oxygen comes from O_2, whose second atom is reduced to water by the electron/hydrogen donor DH_2 (e.g. 2,4-dichlorophenol + $MnCl_2$). In the peroxidase reaction, H_2O_2 accepts electrons or hydrogen from a donor DH_2 which is oxidized to D:

$$\text{peroxidase: } H_2O_2 + DH_2 \rightarrow 2\,H_2O + D.$$

DH_2 may be a broad variety of substrates including phenols, amines, cytochrome c, and also *IAA*. In the case of *IAA* and horseradish peroxidase a free-radical mechanism of *IAA* oxidation has been found [36, 37]. However, the *IAA* free radicals destroy the peroxidase and an autocatalytic oxidation phase then follows the first enzymatic *IAA* peroxidation phase. Although this system might be artificial, it illustrates the complexity of the peroxidative mechanism. It should be noted that both AH_2 and DH_2 are oxidized/dehydrogenated in the reactions, which leaves the possibility open that *IAA* may serve either as an oxygen acceptor or as a hydrogen/electron donor.

Of specific interest here are the effects of *2,4-D* on the peroxidase [34] and on the oxidase [35] isoenzyme spectra. The observed changes differ with the concentration of *2,4-D*, but they are correlated with specific influences of the auxin herbicide on the type of growth. For example, at 0.1–1 μM *2,4-D*, tobacco callus cells grew "tumor-like" and displayed a sharp increase of the content of the rapidly migrating *IAA* oxidase isoenzyme A_5. In contrast, at 10–100 μM *2,4-D*, the A_5 isoenzyme concentration decreased whereas the concentrations of the less rapidly migrating components increased. At these higher concentrations, growth inhibition was observed. The correlation of isoenzyme spectra with growth characteristics has also been observed in morphogenetic mutants of tomato [38] and in the leaves, petioles, and abscission layers of citrus [32]. It has therefore been suggested that the oxidase/peroxidase isoenzyme complement participates in the control of shape and morphogenesis in plants via the regulation of the *IAA* concentration in cells and cell walls [39]. However, some of the isoenzymes will also be concerned with the catalysis of lignification and other processes [32, 40].

IV. Auxins and Nucleic Acid Synthesis

A very pronounced effect after the application of auxin herbicides is the increase in the rate of RNA synthesis and in the RNA concentration [3, 41]. An increased rate of synthesis can be seen a few hours after treatment and has been observed for *2,4-D*, *dicamba*, *picloram*, *IAA*, and other auxin herbicides [2, 42–48]. In the long run, the rate of protein synthesis may also be increased. This second increase is, however, not always observed and, if present, less pronounced than the increase in the rate of RNA synthesis. The result is a decreased protein–to–RNA ratio in plants or tissues treated with auxin herbicides [42]. Furthermore, an antagonism between *2,4-D* and *EPTC* has been explained on this basis in that, whereas *2,4-D* increases the rate of RNA synthesis, *EPTC* decreases the rate of the latter and therefore curbs the harmful consequences of an excessive RNA metabolism.

A more detailed analysis of the increased rate of RNA-synthesis has shown that, while the soluble RNA polymerase II is increased by only 30%, the chromatin-bound RNA polymerase I increases by 800% after application of *2,4-D* [49]. More specifically, the factor γ is doubled in response to the *2,4-D* treatment which activates the nucleolar RNA polymerase I b [50]. This means, that the factor γ appears to be an auxin-induced protein that modulates the activity of RNA polymerase I b. As a consequence, higher amounts of r-RNA (ribosomal RNA) are produced that could give rise to more ribosomes and increased protein synthesis, as long as other factors are not limiting. Some of the current ideas on the action of auxin in plant tissues have been compiled in the following scheme:

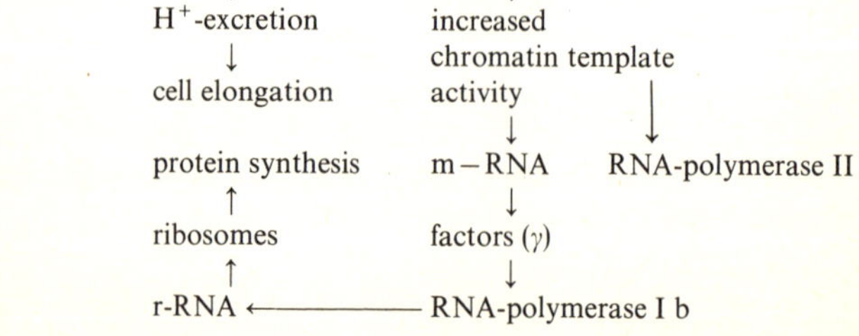

An increased RNA polymerase activity has also been obtained in vitro with preparations of RNA polymerase I from soybean hypocotyls when a *2,4-D* binding protein was added to the assay [51]. It was therefore suggested that the auxin molecule, after originally binding to a component at the plasmalemma, would produce a signal which transmits the stimulus to the chromatin [52]. The nature of this stimulus is unknown, and the possible implications of the binding protein(s) will be discussed separately.

The scheme also indicates that the rapid action of IAA on proton excretion and cell elongation supposedly does not work via the influence on the RNA metabolism. In addition to the RNA polymerase, the activity of a nuclear protein kinase is increased after auxin application [53], and other enzymes or factors might be affected by the hormone in this connection.

V. Auxin-Binding Proteins

Two major concepts of the molecular mode of auxin action have coexisted for a long time and mostly been seen as basically separate responses: (a) induction of proton excretion into the cell-wall compartment followed by increased wall extensibility and tissue elongation and (b) a mechanism of action leading to an increased rate of RNA synthesis. However, it has also been suggested that a single master switch reaction might exist which elicits both responses [4, 4a].

The master switch hypothesis in part rests on the existence of highly active auxin-binding proteins in membrane fractions of maize coleoptile homogenates [4b] (Table 53). Three different membrane fractions have been claimed to display auxin-binding activity. These fractions differ, however, in activity, substrate specificity, and affinity (dissociation constant K_D in Table 53). The highest activity is found in the ER (endoplasmic reticulum) membranes. Here, however, the specificity – as measured by the competition between the binding of ^{14}C-*NAA* and vari-

Table 53. Characterization of auxin-binding proteins; K_D is the dissociation constant in M

Origin of binding protein	Subcellular location(s)	Substrate specificity	K_D	References
Maize coleoptile	Endoplasmatic reticulum (site 1)	Poor	$\sim 10^{-7}$	[55–61]
	Plasma membrane (site 2)	Good	$\sim 10^{-6}$	
	Tonoplast membrane (site 3)	Poor	–	
Maize coleoptile and leaf	Microsomal membranes	Exceptions	4.6×10^{-8}	[62, 63]
Maize coleoptile	Cytosol, soluble protein(s)	Good	–	[64]
Zucchini hypocotyl	Plasma membrane	Good	$\sim 10^{-6}$	[65]
Artichoke tubers	Crude membrane fraction	–	–	[66]
Tobacco pith callus	Particulate fraction	–	3×10^{-7}	[67, 68]
Tobacco pith callus	Cytosol, soluble protein(s)	Good	$10^{-7} - 10^{-8}$	[69, 70]
Soybean cotyledon	Cytosol, soluble protein(s)	–	–	[71]
Pea and bean leaves	Ribulose bisphosphate-carboxylase	Good	$\sim 10^{-6}$	[72, 73]
Bovine serum albumine	–	Good	3.5×10^{-6}	[74, 75]

ous auxins of different potency – is quite poor. On the contrary, the plasmalemma auxin-binding activity displays good to fair substrate specificity when tested with compounds of different auxin activity, i.e. molecules with little or no auxin activity do not compete with ^{14}C-*NAA* for the binding site. However, severe exceptions to this general statement exist [64]. It has been suggested that the ER-site might be involved in auxin-induced proton excretion, which necessitates the transport of ER-vesicles to the plasmalemma for unloading, whereas the plasmalemma site could be involved in the active unidirectional auxin transport [65]. However, after precise reassessment of the *NAA*-binding by maize coleoptile membrane fractions, Murphy [61] concluded that only one binding protein exists.

Successful solubilization with Triton of the membrane-bound auxin binding protein from maize coleoptiles has yielded a molecular weight of 90,000 when inhibitors of protein degradation where present during isolation. Without these inhibitors, proteolytic attack yielded binding proteins with M = 40,000 to 50,000 [63, 76]. Binding occurs only at pH 5.5 and not at pH 7.5 and is saturated at low auxin concentrations. Also, the substrate affinity is within the physiological concentration range. Therefore, though several pieces of evidence are still lacking for the "master switch hypothesis", it appears nevertheless possible that one membrane-integrated auxin-binding protein may elicit both responses mentioned above, proton excretion and increased nucleic acid synthesis. Unfortunately, none of the auxin-binding proteins from maize has displayed any physiological activity so far. However, a *2,4-D*-binding protein isolated from a soybean hypocotyl extract by affinity chromatography did stimulate RNA synthesis in vitro [51].

The concept of a "master switch reaction" envisages a "secondary messenger" which is liberated from the plasma membrane upon binding of an active auxin molecule, and which then diffuses into the cell interior to cause transfer of the signal [77] (see scheme on p. 166). The nature of this anticipated "secondary messenger" is not known. The auxin-binding protein is "neither itself a proton-pumping ATPase nor a separate entity that modulates the activity of such an ATPase" [76]. However, very recently, an auxin stimulated ATPase has been reported [76a].

Auxin-binding proteins have also been found in other plant species and plant tissues, although the binding observed there was always considerably lower as compared with the maize coleoptile (Table 53). Also, in addition to membraneous auxin-binding proteins, soluble proteins have been found with good and specific auxin-binding characteristics. An interesting case is the specific auxin binding by ribulose bisphosphate carboxylase, which has been suggested to have a function in auxin transport, sequestration and/or light modulation. The carboxylase auxin-binding activity is lost in the highly purified enzyme; this finding again suggests the existence of highly specific soluble auxin-binding proteins. The soluble binding protein found in tobacco pith callus is present only in certain growth phases. Also, the existence of endogenous factors, which modify the affinity of the auxin-binding proteins, has been reported [59, 63, 78].

Though auxin-binding proteins with appropriate substrate specificities undoubtedly exist in higher plant cells, their relevance is not clear at all. Firstly, none of the membraneous auxin-binding proteins has exhibited any physiological activity so far. And, most importantly, auxin binding by the lipophilic protein bovine serum albumine (BSA) has been shown to display substrate specificities and binding affinities largely similar to those of the extensively investigated auxin-binding proteins of plant origin [75]. A BSA-sepharose has even been used as an affinity column material for the isolation of IAA from plant extracts [79]. The possibility should therefore be seen that some lipophilic plant proteins bind small molecules by unspecific hydrophobic interaction which happens to include some active auxins [61, 64]. Consequently, as long as a physiological role for the auxin-binding proteins has not unequivocally been shown, any further conclusions should be treated with caution.

VI. Sublethal Effects on Intact Growing Plants

Treatment of plants with auxin-active herbicides may result in stimulations of biochemical pathways, but also in stimulations and perturbations of growth [80]. The stimulation of RNA synthesis has already been discussed (p. 166). In addition, also the rate of lipid synthesis as measured by the incorporation of ^{14}C-malonate [81] or ^{14}C-acetate [82] has been found to be increased. The insoluble carbohydrate reserves (starch) appear to be mobilized which results in increased levels of free soluble carbohydrates [83]. The level of total amino acids may also be increased [84] and, depending on the experimental conditions, the amino acid spectrum may be changed [84] or unchanged [85].

Since the auxin sensitivity of a tissue strongly depends on its physiological stage, the responses described above may vary quite considerably and may either occur not at all or, in an advanced stage of action, turn into the opposite. This, however, means herbicidal rather than sublethal action.

An indication of the occurrence of sublethal biochemical plant responses to auxin herbicides in the field may be seen in reports on changed plant susceptibility to attack by certain pests. The increased sensitivity to "high-sugar deseases" and the decreased sensitivity to "low-sugar deseases" may be correlated with the responses of the soluble carbohydrate levels [86] (cf. also p. 76). Similarly, an increased attack by aphids has been ascribed to increased protein levels and possibly also to soluble amino acids in *2,4-D* treated plants [87]. Compositional changes, which have been correlated with a state of senescence, have recently also been reported for the chloroplast pigments in *MCPA*-treated barley plants [88].

At the morphological level, epinasty – i.e. a faster growth rate of the upper tissue side of organs like leaves, stems, and petioles resulting in a downward curling response –, thickening of conductive tissue and leaf growth anomalies are generally observed [89]. These responses are basically the same for sublethal and herbicidal concentrations, although the extent varies accordingly. It appears, therefore, that sublethal effects, after treatment of intact plants with auxin herbicides, are merely subherbicidal responses and are mechanistically closely related to the herbicidal response.

VII. Effects on Basic Plant Metabolism

Indirect effects of auxin-active herbicides on plant metabolic pathways like nucleic acid, protein and lipid synthesis, respiration and photosynthesis are very common and have already been discussed to some extent. On the other hand, a direct influence on the metabolic reactions involved in the above mentioned pathways may only occur at high concentrations. In general, between 0.1 and 10 mM of auxin-type herbicides are required to cause inhibition of photosynthesis and respiration when the effect on these processes is measured in vitro [90, 91]. Among the phenoxy herbicides, one of the strongest inhibitors of the Hill reaction in isolated chloroplasts is *2,4,5-T* with a pI_{50}-value of 3.6 which still is very low (cf. Tables 4–15). In a comparison of the action of 6 phenoxy compounds on isolated mitochondria, Matlib et al. [92] generally found uncoupling activity in the 0.1–1 mM concentration range whereas 1–4 mM were required for the inhibition of respiration. In in-vivo measurements of the ATP level after herbicidal treatment of soybean hypocotyls, no deviations from the control value were found for 0.2–0.6 mM *dicamba* [93]. On the other hand, mitochondria isolated from soybean seedlings which had been pretreated with *2,4-D* displayed increased activities [94]. Another indirect effect is the decreased PAL (phenylalanine ammonia lyase) formation in *picloram*-treated tobacco [95]. Since PAL is the key enzyme for the regulation of the phenylpropanoid metabolism (p. 185) (lignin, anthocyanins), any changes here may have profound regulatory consequences.

Some controversy concerning the effect of phenoxy herbicides on mitotic reactions exists. Although mitotic aberrations have occasionally been observed to be induced after tretment of plants with *2,4-D*, these may largely depend on concentration. In *Daucus carota* cell-suspension cultures, 0.1–20 $\mu g\, ml^{-1}$ of *2,4-D* progressively increased the duration of the cell cycle from 50 to 250 h, mainly by blockage of the G 1 or G 2 stages of the cell cycle [96]. At $>30\ \mu g\, ml^{-1}$ *2,4-D* spindle formation might be prevented. Similarly, effects of *2,4-D* on spindle formation and chromosome morphology reported by other authors occurred at concentrations between 10 and 100 $\mu g\, ml^{-1}$ [97–99]. For an inhibition of DNA synthesis in chick cell cultures 600 $\mu g\, ml^{-}1$ *2,4-D* were required [100]. On the other hand, at 0.1–10 $\mu g\, ml^{-1}$ *2,4-D* no mitotic irregularities were found with different plant tissue cultures [101]. Ehrenberg [102] therefore concluded that "the genetic risk, if it exists, is low". These data again illustrate the great importance of the concentration used in a particular study (p. 7).

VIII. Selectivity

Resistance against the herbicidal action of auxin herbicides may arise from three different mechanisms, which include (a) differences in the rates of metabolic inactivation, (b) differences in the rates of uptake and/or transport and (c) differences in the physiology between sensitive and resistant plants. Whereas most auxin herbicides are rapidly metabolized in plant tissues, and differences in the rates of uptake and transport have been observed among sensitive and resistant plants, the main reason for the resistance against auxin herbicides appears to reside in differences of the specific physiology of the plants.

Metabolism. Much work has been done on the metabolism of auxins and auxin herbicides in plant tissues and cell cultures, and only some aspects relevant for the discussion of selectivity are repeated here. Whereas the endogenous auxin *IAA* may mainly be inactivated by oxidation and/or peroxidation (p. 165), the excessive concentrations of externally applied artificial auxins react with the amino acids aspartate and glutamate and form the respective amides (see for *2,4-D* [103–110, 141], for *2,4-DB* [111–112], for *MCPA* [113], for *2,4,5-T* [114], for *picloram* [115, 116]). However, *IAA* may also be complexed with aspartate and glutamate [69]. In addition, hydroxylation of the ring and subsequent glycosidation contribute to the inactivation of these herbicides [117]. It appears, that lower plants such as mosses tend to form glucose complexes while the flowering plants tend to produce aspartic or glutamic acid complexes, the aspartic acid complexes being clearly more important [5]. In any case, only the excess of the hormone appears to be conjugated with either glucose or an amino acid, and a free pool of 1 to 4 n moles per g fresh weight of the treated plants remains in the tissue [107].

As mentioned before, a free carboxy group and a certain distance from the ring are prerequisites to an auxin-active molecule (p. 161). Auxin herbicides not meeting these requirements must be converted into the active molecule. This has e.g. been shown for *2,4-DB* where β-oxidation yields *2,4-D* [118, 119], and for *orthonil* which is converted into the respective free acid [20].

The following scheme illustrates the possible implications of auxin metabolism with respect to auxin action:

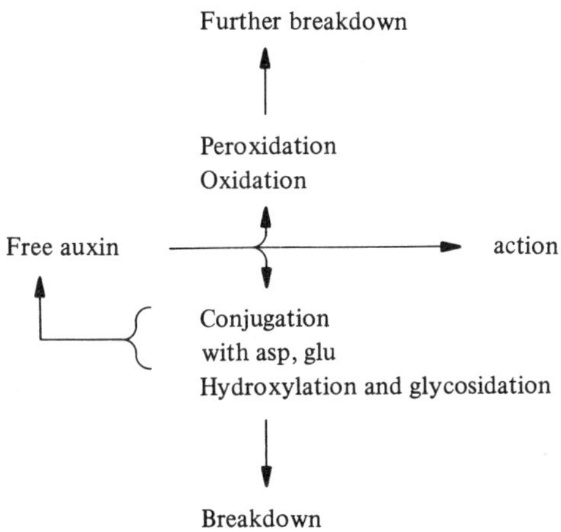

Since the amino acid conjugates of auxin herbicides have clearly been shown to be physiologically active when applied to a sensitive tissue, they might also be considered as storage forms [108, 116]. In addition, Venis [120] has reported that only physiologically active auxins induce an increase of the enzyme N-acylaspartate synthetase. The possibility should therefore be considered that the physiological/biochemical primary auxin action produces a feed-back signal which controls the concentration of the free auxin. The suggestion of three interdependent auxin pools:

$$\text{transport pool} \leftrightarrows \text{free pool} \rightleftharpoons \text{immobilized pool},$$

would then equally apply to endogenous and to herbicidal auxins.

Transport. The uptake and translocation of phenoxy acids applied to foliage in sensitive and resistant tissues has been summarized in detail in an earlier paper [121]. It has been stated there that the transport of these herbicides would occur in the phloem and that the differential activity may be attributed at least in part to differential movement, but also to differences occurring during the uptake into the phloem. Also, physiological conditions and the physiological effects of the herbicide at the application site, e.g. the creation of metabolic sinks, tend to greatly influence the mobility. The transport of *chloramben* into the shoot was reported to be eight times higher in sensitive *Echinochloa* as compared with resistant wheat [122]. However, wheat glucosylates the herbicide in the root, which may be the basis for the lack of mobility. Another review on foliar absorption and translocation has appeared more recently [123]. Here again, the great influence of external factors was stressed.

Physiology. Though the available knowledge on uptake, transport and metabolism of auxin herbicides clearly shows that the physiological activity might be greatly influenced by these parameters, a general correlation with resistance is not obvious. On the other hand, resistance of plants to auxin herbicides has shown good correlations with the nuclear volume of the respective plant species which, in turn, is positively correlated with chromosome number, polyploidization and perennial growth habit [124, 125]. Furthermore, plants resistant to auxin herbicides contain higher nuclease levels (*2,4-D* [3]; *picloram* [45]). Sensitive plants therefore respond to the application of auxin herbicides by an increase of the ratio RNA/protein whereas for resistant plants the reverse is true [42].

It is at present not possible to estimate the contribution of the different physiological peculiarities in resistant plants. However, it appears reasonable to advance the hypothesis that plants resistant to auxin-herbicide actions are resistant because of peculiar physiological growth characteristics which clearly distinguish them from the sensitive plants. The high ribonuclease levels in gramineous species are a good example for the illustration of this hypothesis, but other biochemical, physiological, and anatomical characteristics in resistant plants will presumably also contribute to the resistance.

The selection of resistant strains has been possible at the level of intact plants [126–128], but also with cell culture systems [129, 130]. Genotypic variation has been the basis for the selection in the field.

IX. Phytotoxicity

In summarizing the available information and relating it to the sequence of effects leading from herbicide application to herbicidal action, three phases in the response of the plant may be discerned [3, 131]:

phase 1: 0–2 d: stimulation: increase of photosynthesis, ion uptake, RNA, weight, etc.; reserve mobilization
phase 2: 2–7 d: distribution: stem elongation, tissue swelling, leaf decay; curling
phase 3: 7–10 d: collaps: tissue decay.

When looking for the actual cause of the herbicidal activity, the aberrant nucleic acid metabolism should be primarily considered. This aberrant nucleic acid metabolism is in turn caused by the hormonal imbalance in the treated tissue. Then, in the course of the increased production of nucleic acids, many biochemical processes are stimulated as indicated above in "phase 1". During this phase, cell division is promoted and new cambial primordia are created [132].

Although the stimulation of cell divisions continues into "phase 2", cell growth and tissue swelling become more important with time and lead to the aberrant morphology commonly seen during the action of auxin herbicides. The natural source – sink relationships are altered, and the translocation of assimilates is consequently severely affected [121]. Eventually, the interplay and cooperation of the different tissues constituting the plant breaks down completely and the tissue decays.

Three phases of auxin-herbicide action have also been discerned at the subcellular level: tonoplast disintegration during phase 1, chloroplast grana swelling during phase 2 and plasmalemma disruption with complete intracellular breakdown during phase 3 [133]. The early damage of the chloroplast membranes during the course of the herbicidal action of auxins has repeatedly been observed and also been correlated with an accumulation of *2,4-D* in the chloroplast [134–136]. On the other hand, although *picloram* affects primarily the chloroplast [134], it does not accumulate in this organelle [137]. Therefore, although the chloroplast appears to be damaged early during the course of herbicidal action of artificial auxins, it should not be considered as a "site of herbicidal action" in the sense that molecular auxin interference at the chloroplast level is implied. The effects of auxin herbicides at the chloroplast level could equally well be a consequence of molecular auxin interferences at other subcellular sites. Any conclusions as to the molecular site of action drawn from subcellular distribution or toxicity studies suffer from the possibility that the sites of action, accumulation and primary damage may all be different [138].

References for I

1. Abeles, F.B.: Ethylene in plant biology, p. 70. London, New York, San Francisco: Academic Press 1973
2. Cherry, J.H.: Hort. Sci. *5*, 205 (1970)
3. Hanson, J.B., Slife, F.W.: Resid. Rev. *25*, 59 (1969)
4. Morré, D.J., Cherry, J.H.: Auxin hormone-plasma membrane interactions. In: Plant growth regulation. Pilet, P.E. (ed.), p. 35. Berlin, Heidelberg, New York: Springer 1977

4a. Vanderhoef, L.N., Stahl: Proc. Natl. Acad. Sci. U.S.A. *72*, 1822 (1975)

4b. Stoddard, J., Venis, M.A.: Encycl. Plant Physiol. *9*, 445 (1980)

5. Thimann, K.V.: Hormone action in the whole life of plants, p. 207. Amherst: The Univ. of Mass. Press 1977
6. Council for Agricultural Sci. and Technol. (Cast): Weed Sci. *23*, 253 (1975)
7. Masuda, Y.: Planta *83*, 171 (1968)
8. Hager, A., Menzel, H., Krauss, A.: Planta *100*, 47 (1971)
9. Evans, M.L., Vesper, M.J.: Plant Physiol. *66*, 561 (1980)
10. Vanderhoef, L.N. et al.: Plant Physiol. *59*, 1000 (1977)
11. Vanderhoef, L.N., Lu, T.-Y.S., Williams, C.A.: Plant Physiol. *59*, 1004 (1977)
12. Rayle, D.L., Cleland, R.E.: Plant Physiol. *66*, 433 (1980)
13. Vendrig, J.: Nature *234*, 557 (1971)
14. Ingram, D.S., Butcher, D.N.: Z. Pflanzenphysiol. *66*, 206 (1972)
15. Bayliss, M.W.: Plant Sci. Lett. *8*, 99 (1977)
16. Lee, Q.P.: Weed Abstr. *23*, no. 1334 (1974)
17. Diaz-Colon, J.D. et al.: Physiol. Plant. *27*, 60 (1972)
18. Katekar, G.F., Geissler, A.E.: Plant Physiol. *60*, 826 (1977)

19. Mitchell, J.W., Livingston, G.A.: Methods of studying plant hormones and growth-regulating substances, pp. 140, Agriculture Handbook No. 336. Washington, United States Department of Agriculture 1968
20. Vendrig, J., Dierickx, P.: Arch. of Environ. Contam. and Toxicol. *4*, 395 (1976)
21. Farrimond, J.A., Elliott, M.C., Clack, D.W.: Phytochem. *19*, 367 (1980)
22. Katekar, G.F.: Phytochem. *18*, 223 (1979)
22a. Farrimond, J.A., Elliott, M.C., Clack, D.W.: Phytochem. *20*, 1185 (1981)
23. Kaethner, T.M.: Nature *267*, 19 (1977)
24. Skoog, F., Miller, C.O.: Symp. Soc. Exp. Biol. *11*, 118 (1957)
25. Skoog, F., Armstrong, D.J.: Annu. Rev. Plant Physiol. *21*, 359 (1970)
26. Pratt, H.K., Goeschl, J.D.: Annu. Rev. Plant Physiol. *20*, 541 (1969)
27. Stacewicz-Sapuncakis, M. et al.: Plant Physiol. *52*, 466 (1973)
28. Morgan, P.W., Baur, J.R.: Plant Physiol. *46*, 655 (1970)
29. Scott, P.C., Morris, R.O.: Nature *227*, 1366 (1970)
30. Holm, R.E. et al.: Plant Physiol. *45*, 41 (1970)
31. Hallaway, M., Osborne, D.J.: Science *163*, 1067 (1969)
32. Gaspar, T. et al.: Plant Cell Envir. *1*, 225 (1978)
33. Arnison, P.G., Boll, W.G.: Canad. J. Bot. *54*, 1847 (1976)
34. Arnison, P.G., Boll, W.G.: Canad. J. Bot. *54*, 1857 (1976)
35. Lee, T.T.: Plant Physiol. *49*, 957 (1972)
36. Fox, L.R., Purves, W.K., Nakada, H.J.: Biochemistry *4*, 2754 (1965)
37. Fox, L.R., Purves, W.K.: The mechanism of peroxidase catalysis of 3-indoleacetic acid oxidation. In: Biochemistry and physiology of plant growth substances. Wightmann, F., Setterfield, G. (eds.), p. 301. Ottawa: The Runge Press 1968
38. Palmieri, S. et al.: Physiol. Plant. *42*, 85 (1978)
39. Bryant, S.D., Lane, F.E.: Plant Physiol. *63*, 696 (1979)
40. Scandalious, J.G., Sorenson, J.C.: Isoenzymes in plant tissue culture. In: Plant cell, tissue, and organ culture. Reinert, J., Bajaj, Y.P.S. (eds.), p. 719. Berlin, Heidelberg, New York: Springer 1977
41. Cherry, J.H.: Actions on nucleic acid and protein metabolism. In: Herbicides – physiology, biochemistry, ecology, Vol. I. Audus, L.J. (ed.), p. 525. London, New York, San Francisco: Academic Press 1976[2]
42. Chen, L.G., Switzer, C.M., Fletcher, R.A.: Weed Sci. *20*, 53 (1972)
43. Chen, L.G. et al.: Weed Sci. *21*, 181 (1973)
44. Cherry, J.H. et al.: Plant Physiol. *48*, 43 (1971)
45. Malhotra, S.S., Hanson, J.B.: Weed Sci. *18*, 1 (1970)
46. Arnold, W.E., Nalewaja, J.D.: Weed Sci. *19*, 301 (1971)
47. Trewavas, A.: Phytochem. *7*, 673 (1968)
48. Sharma, M.P., van den Born, W.H.: Canad. J. Bot. *50*, 2039 (1972)
49. Guilfoyle, T.J. et al.: Proc. Natl. Acad. Sci. U.S.A. *72*, 69 (1975)
50. Teissere, M. et al.: Biochim. Biophys. Acta *402*, 391 (1975)
51. Rizzo, P.J., Pedersen, K., Cherry, J.H.: Plant Sci. Lett. *8*, 205 (1977)
52. Jacobsen, J.V.: Annu. Rev. Plant Physiol. *28*, 537 (1977)
53. Murray, M.G., Key, J.L.: Plant Physiol. *61*, 190 (1978)
55. Batt, S.: Planta *130*, 15 (1976)
56. Batt, S., Wilkins, M.B.: Planta *130*, 7 (1976)
57. Dohrmann, U., Hertel, R., Kowalik, H.: Planta *140*, 97 (1978)
58. Ray, P.M.: Plant Physiol. *59*, 594 (1977)
59. Ray, P.M., Dohrmann, U., Hertel, R.: Plant Physiol. *59*, 357 (1977)
60. Ray, P.M., Dohrmann, U., Hertel, R.: Plant Physiol. *60*, 585 (1977)
61. Murphy, G.J.P.: Planta *149*, 417 (1980)
62. Cross, J.W. et al.: Plant Physiol. *61*, 581 (1978)
63. Cross, J.W., Briggs, W.R.: Plant Physiol. *62*, 152 (1978)
64. Murphy, G.J.P.: Plant Sci. Lett. *19*, 157 (1980)
65. Jacobs, M., Hertel, R.: Planta *142*, 1 (1978)
66. Trewavas, A.: Phytochem. *19*, 1303 (1980)
67. Vreugdenhil, D., Burgers, A., Libbenga, K.R.: Plant Sci. Lett. *16*, 115 (1979)
68. Vreugdenhil, D., Harkes, P.A.A., Libbenga, K.R.: Planta *150*, 9 (1980)

69. Bogers, R.J. et al.: Plant Sci. Lett. *19*, 311 (1980)
70. Oostrom, H. et al.: Planta *149*, 44 (1980)
71. Ihl, M.: Planta *131*, 223 (1976)
72. Wardrop, A.J., Polya, G.M.: Plant Physiol. *66*, 105 (1980)
73. Wardrop, A.J., Polya, G.M.: Plant Physiol. *66*, 112 (1980)
74. Murphy, G.J.P.: Pestic. Sci. *9*, 161 (1978)
75. Murphy, G.J.P.: Plant Sci. Lett. *15*, 183 (1979)
76. Lamb, C.J.: Nature *274*, 312 (1978)
76a. Scherer, G.F.E.: Planta *151*, 434 (1981)
77. Rodbell, M.: Nature *284*, 17 (1980)
78. Venis, M.A., Watson, P.J.: Planta *142*, 103 (1978)
79. Schulze, A., Bandurski, R.S.: Plant Physiol. Suppl. *63*, no. 277 (1979)
80. Ries, S.K.: Subtoxic effects on plants. In: Herbicides-physiology, biochemistry, ecology, Vol. 2. Audus, L.J. (ed.), p. 313. London, New York, San Francisco: Academic Press 1976[2]
81. Mann, J.D., Pu, M.: Weed Sci. *16*, 197 (1968)
82. Muslih, R.K., Linscott, D.L.: Plant Physiol. *60*, 730 (1977)
83. Irvine, A.M., Forbes, J.C., Draper, S.R.: Weed Res. *17*, 169 (1977)
84. Jung, K., Blohm, D.: Biolog. Zentralblatt *93*, 571 (1974)
85. Pellet, P.L., Saghir, A.R.: Weed Res. *11*, 182 (1971)
86. Horsfall, J.G., Dimond, A.E.: Z. PflKrankh. PflSchutz *64*, 415 (1957)
87. Pimentel, D.: Science *193*, 239 (1976)
88. Kleudgen, H.K.: Z. Naturforsch. *34c*, 106 (1979)
89. Martin, J.A., Fletcher, J.T.: Weed Res. *12*, 268 (1972)
90. Robertson, M.M., Kirkwood, R.C.: Weed Res. *10*, 94 (1970)
91. Diaz, M.A., Chueca, A., Gorge, J.L.: Pestic. Biochem. Physiol. *13*, 105 (1980)
92. Matlib, M.A. et al.: J. Exp. Bot. *23*, 886 (1972)
93. Gruenhagen, R.D., Moreland, D.E.: Weeds *19*, 319 (1971)
94. Baxter, R., Hanson, J.B.: Planta *82*, 246 (1968)
95. Paynot, M. et al.: Phytochem. *15*, 669 (1976)
96. Bayliss, M.W.: Plant Sci. Lett. *8*, 99 (1977)
97. Sen, S.: Naturwiss. *62*, 184 (1975)
98. Sen, S.: Experientia *34*, 724 (1978)
99. Ganesh, P.: Weed Abstr. *27*, no. 4209 (1978)
100. Haag, D., Goerttler, K., Preiss, D.: Arch. Toxicol. *33*, 91 (1975)
101. Singh, B.D., Harvey, B.L.: Experientia *31*, 785 (1975)
102. Ehrenberg, L.: Weed Abstr. *28*, no. 1384 (1979)
103. Feung, C.S. et al.: J. Agric. Food Chem. *19*, 475 (1971)
104. Feung, C.S. et al.: Plant Physiol. *50*, 80 (1972)
105. Feung, C.S. et al.: J. Agric. Food Chem. *23*, 373 (1975)
106. Feung, C.S. et al.: J. Agric. Food Chem. *26*, 1064 (1978)
107. Davidonis, G.H. et al.: Plant Physiol. *62*, 80 (1978)
108. Davidonis, G.H. et al.: J. Agric. Food Chem. *27*, 1086 (1979)
109. Davidonis, G.H. et al.: Plant Physiol. *65*, 94 (1980)
110. Davidonis, G.H. et al.: Plant Physiol. *66*, 537 (1980)
111. Smith, A.E., Oswald, T.H.: Weed Sci. *27*, 389 (1979)
112. Smith, A.E.: Weed Sci. *27*, 392 (1979)
113. Collins, D.J., Gaunt, J.K.: Biochem. J. *124*, 9P (1971)
114. Arjmand, M., Hamilton, R.H., Mumma, R.O.: J. Agric. Food Chem. *26*, 1125 (1978)
115. Scott, P.C., Morris, R.O.: Plant Physiol. *46*, 680 (1970)
116. Hallmen, U.: Physiol. Plant. *34*, 266 (1975)
117. Hamilton, R.H. et al.: J. Agric. Food Chem. *19*, 480 (1971)
118. Garraway, J.L.: Pestic. Sci. *1*, 132 (1970)
119. McComb, A.J., McComb, J.A.: Plant Sci. Lett. *11*, 227 (1978)
120. Venis, A.: Plant Physiol. *49*, 24 (1972)
121. Robertson, M.M., Kirkwood, R.C.: Weed Res. *10*, 94 (1970)
122. Stoller, E.W.: Plant Physiol. *46*, 732 (1970)
123. Richardson, R.G.: Weed Res. *17*, 259 (1977)

124. Mohandas, T., Grant, W.F.: Canad. J. Bot. *51*, 1133 (1973)
125. Tomkins, G.M., Grant, W.F.: Canad. J. Bot. *56*, 1466 (1978)
126. Ellis, M., Kay, Q.O.N.: Weed Res. *15*, 307 (1975)
127. Ellis, M., Kay, Q.O.N.: Weed Res. *15*, 317 (1975)
128. Ellis, M., Kay, Q.O.N.: Weed. Res. *15*, 327 (1975)
129. Oswald, T.H.: Canad. J. Bot. *55*, 1351 (1977)
130. Chaleff, R.S., Parsons, M.F.: Proc. Natl. Acad. Sci. U.S.A. *75*, 5104 (1978)
131. Cardenas, J. et al.: Weed Sci. *16*, 96 (1968)
132. Rogerson, A.B. et al.: Weed Sci. *20*, 445 (1972)
133. White, J.A., Hemphill, D.D.: Weed Sci. *20*, 478 (1972)
134. Ayling, R.D.: Weed Res. *16*, 301 (1976)
135. Hallam, N.D., Sargent, J.A.: Planta *94*, 291 (1970)
136. Hallam, N.D.: J. Exp. Bot. *21*, 1031 (1970)
137. Baur, J.R., Bowman, J.J.: Physiol. Plant. *28*, 372 (1973)
138. Anderson, J.L., Thomson, W.W.: Resid. Rev. *47*, 167 (1973)
139. Cleland, R.E.: Auxin and H^+-excretion: The state of our knowledge. In: Plant growth substances. Skoog, F. (ed.), p. 71. Berlin, Heidelberg, New York: Springer 1980
140. Montague, M.J. et al.: Plant Physiol. *67*, 701 (1981)
141. Montague, M.J. et al.: Plant Physiol. *67*, 603 (1981)
142. Trewavas, A.: Plant Cell Envir. *4*, 203 (1981)

K. Auxin-Inhibitor Herbicides

I. Structural Requirements

The group of herbicides discussed under the heading "auxin inhibitors" was developed originally for the control of wild oats (*Avena fatua* L.). This species became a serious problem in small grain crops after the introduction of the phenoxyalkanoic acid herbicides which generally do not control gramineous species (p. 172). Therefore, over the past 30 years wild oats has spread and infested large areas of agricultural production.

It has become increasingly clear that these herbicides developed for wild oat control are not auxins themselves like phenoxy acid herbicides but they interfere with the action of the endogenous auxin, *IAA*. Auxins contain a free carboxy group in their molecule, and it may be expected that compounds that interfere with the action of auxins should also contain a free carboxy group. Therefore, although the herbicides listed in Table 54 are all esters, it is very likely, according to most experimental evidence, that the biologically active molecules are the free acids. Table 54 includes also compounds that are known to be auxin-transport inhibitors *(DPX 1840, naptalam, TIBA)* but are not herbicidally active. These compounds are relevant to the discussion of the effects of group 1 herbicides on auxin-dependent systems. With *DPX 1840*, a conversion to the free acid is required for biological activity [1]. In the case of *difenopenten* activation of the herbicide by *β*-oxidation may be expected.

The herbicidal compounds (group 1) listed in Table 54 contain an asymmetric C-atom adjacent to the carboxyl group. In two cases where the enantiomers have been tested separately only the l-enantiomer was found to be biologically active (see *chlorfenprop-methyl* [2, 3] and *flamprop-isobutyl* [4]) (Table 55). The active molecule of *chlorfenprop-methyl* has the absolute configuration of R, i.e. *D-(−)-chlorfenprop-metyhl* [5]. This specific form was synthesized by the side-specific hydrogenation of the respective cinnamic acid with enzymes from the anaerobic bacterium, *Clostridium klujveri*. The third successful preparation of the individual enantiomers has been reported for *diclofop-methyl* [47]. In this case, again the R-configuration was found to be the active molecule, more precisely D(R)-*diclofop-methyl.*

The structure-activity relationships for auxin-transport inhibitors [6] may be applicable to the herbicides of group 1 when considering them as "auxin-transport inhibitors". Clearly, a carboxy group and one or two planar ring systems characteristic of auxin transport inhibitors are present in the active herbicide molecules, but the required distance of 7–8 Å between two aromatic substituents does not ex-

Table 54. Herbicides (group 1) and some other structures (group 2) which inhibit auxin action; asymmetric C-atom denoted by *C

Common name or equivalent	Structure	Water sol. ($mg\ l^{-1}$)
Group 1:		
Benzoylprop-ethyl	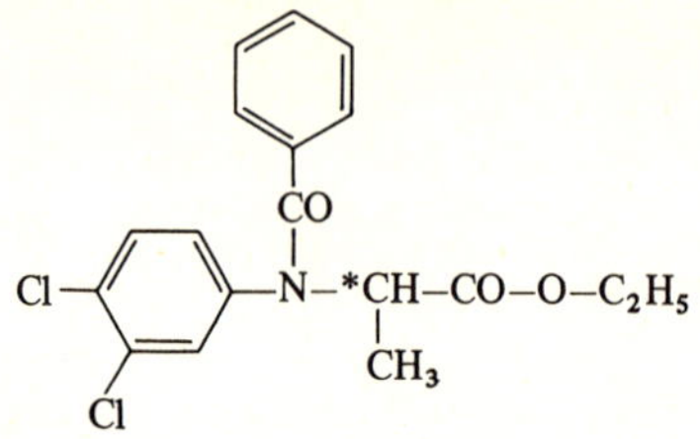	20
Chlorfenprop-methyl	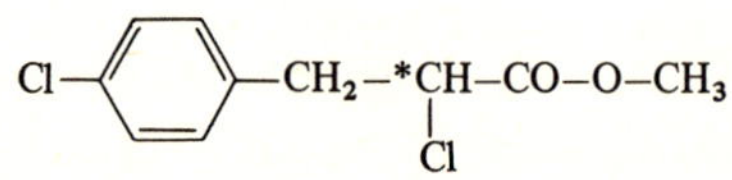	40
Clofop-isobutyl	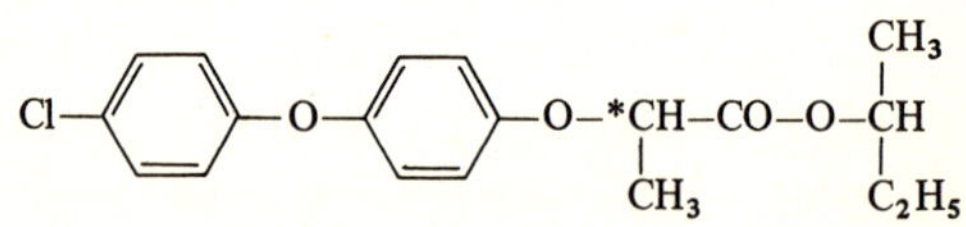	180
Diclofop-methyl	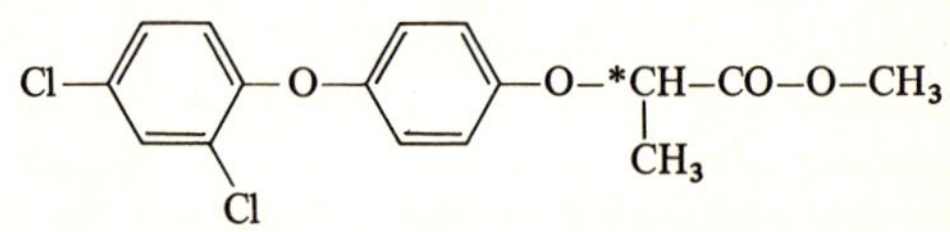	0.3 [50]
Trifop-methyl	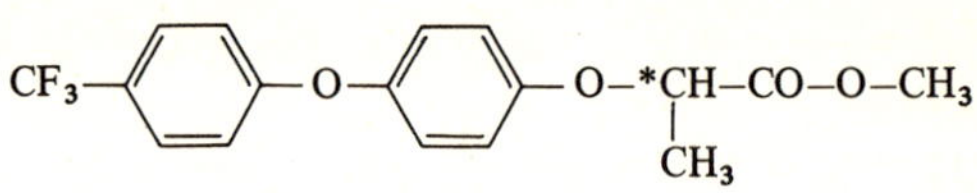	–
Difenopenten	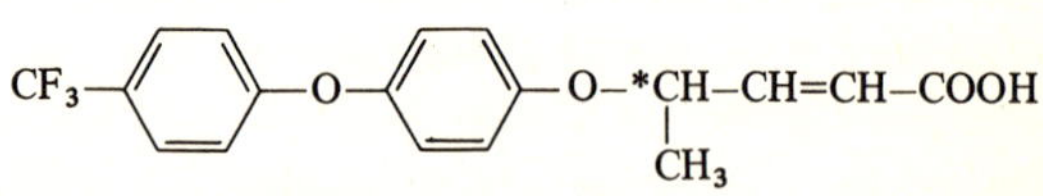	–
Flamprop-isopropyl	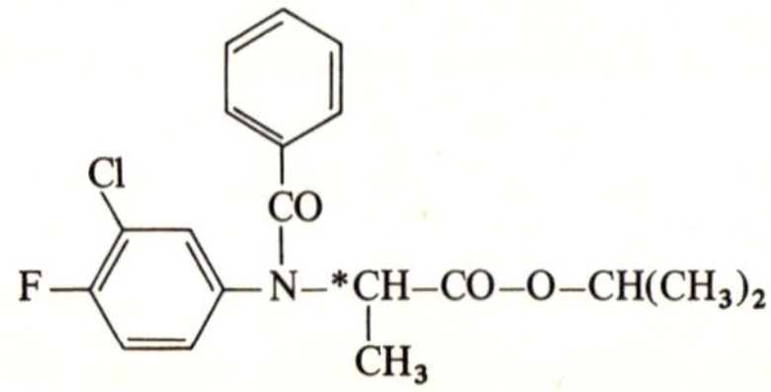	18
Flamprop-methyl	—— // —— —O–CH_3	35

Table 54 (continued)

Common name or equivalent	Structure	Water sol. (mg l^{-1})
Group 2:		
DPX-1840	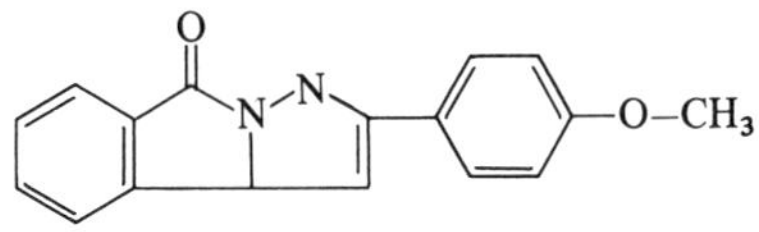	–
Naptalam=NPA	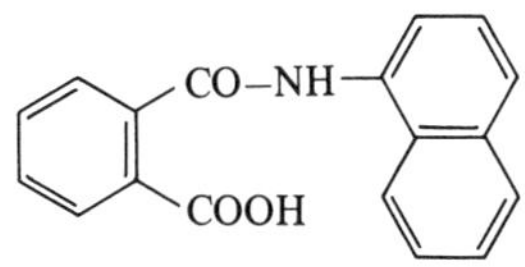	200 (Na-salt: 30%)
TIBA	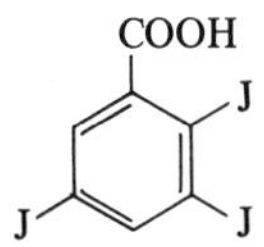	–

Table 55. Activity and selectivity of chiral forms of *flamprop-isopropyl* [4]. GID_{50}, GID_{90}: growth inhibition dose for 50% or 90% inhibition, respectively

	Oat GID_{90} (kg/ha)	Barley GID_{50} (kg/ha)
Racemate	0.84	3.63
l-(−)enantiomer	0.44	2.94
d-(+)enantiomer	10.23	4.17

ist in the herbicides. However, there are other exceptions to the suggested rules such as *TIBA*. The typical auxin transport inhibitors are also not herbicidally active. The authors therefore discuss several different receptor sites for different classes of auxin-transport inhibitors, and the herbicides in Table 54 should certainly be considered in this context.

II. Degradation and Selectivity

Although the herbicidal activity appears to be due to the formation of the free acid in treated tissues, the basis for the selectivity is not the same between the herbicides in this series. Two groups may be distinguished; one (a) where the selectivity is related to the rate of degradation and a second (b) where degradation is apparently not the primary selective mechanism. However, with both groups hydrolysis, subsequent metabolism, herbicidal activity, and selectivity are interrelated and can only be understood together.

(a) *Benzoylprop-ethyl*, *Flamprop-isobutyl*, and *Flamprop-methyl*

The in vivo hydrolysis rate was higher in sensitive oats than in resistant wheat or barley for the herbicides *benzoylprop-ethyl* [7,8], *flamprop-isobutyl* [9], and *flamprop-methyl* [10]. However, not only the rate of hydrolysis but also the rate of subsequent metabolism of the herbicidally active free acid may be important in selectivity [8, 11–13]. The further metabolism of the acids, *benzoylprop* and *flamprop*, occurs mainly via hydroxylation and conjugation; hydroxylation of the parent ester is not a significant reaction in the overall metabolism. Selectivity may also partly depend on the mobilities of *benzoylprop-ethyl* and the corresponding acid, *benzoylprop;* the ester is relatively immobile while the acid is relatively mobile in the plant [14].

An esterase that hydrolyzes *benzoylprop-ethyl* ($k_m = 3.8$ μM) and also *diclofop-methyl* has been partially purified from wild oats [7]. Extracts from wheat contained an inhibitor that inhibited the esterase from both wild oats and wheat. Evidently, the esterase from wheat was only observed after its separation from the inhibitor.

(b) *Chlorfenprop-methyl*, *Diclofop-methyl*, *Clofop-isobutyl*, and *Trifop-methyl*

Data on these herbicides indicate no significant differences between sensitive wild oat or sensitive oat varieties and several resistant species with respect to the rates of uptake, transport, and degradation [15–21]. However, the degradation pathway for the subsequent metabolism of the acid, *diclofop*, differed greatly between sensitive wild oat and resistant wheat [17, 18, 22, 23].

Two theories, therefore, exist at present to account for the selective herbicidal action of these herbicides: (a) either the site of herbicidal binding and action differs between sensitive and resistant species, or (b) the differences in subsequent metabolism of the acid provide a basis for selectivity. Whereas the first alternative has been suggested on the basis of studies with closely related *Avena* varieties [15] and of levels of the free acid in sensitive and resistant species [20], the second alternative is equally possible because actual detoxification of *diclofop* by ring hydroxylation and subsequent complex formation occurs predominantly in resistant plants [17,18]. Differences in the pathway of further metabolism have also been reported for *chlorfenprop* [24]. In the case of *diclofop* the herbicide-susceptible wild oat predominantly conjugates the free acid to an ester conjugate; this conjugation may be reversible and could regenerate the herbicidally active *diclofop* over a prolonged time, thus rendering the plant susceptible to this compound.

A peculiarity of several compounds in this series is the relatively low activity of the acid as compared to the ester when applied to intact tissue pieces (for *chlorfenprop-methyl* [3,15], for *DPX 1840* [1], for *diclofop-methyl* [25]). However, the herbicidal activity of *chlorfenprop*, the free acid, was equal to that of the ester when applied to intact plants. For an understanding of these differences it should be recalled that experiments with tissue pieces usually require a few minutes up to a few hours, whereas in experiments using intact plants the herbicidal symptoms are observed over a longer period. It was concluded from these observations, that the site of molecular interaction may be situated in a lipophilic compartment which may only slowly be penetrated by the active acid [15]. The ester, on the contrary, may penetrate the tissue easily and may be hydrolyzed in close proximity to the site of action.

III. Inhibition of Auxin-Dependent Systems

Two main pieces of evidence support the view that the herbicides listed in Table 54 are "auxin inhibitors": (a) the rapid and effective inhibition of auxin-dependent responses and (b) the antagonistic interaction with auxin herbicides, e.g. of the phenoxyacetic acid and the benzoic acid type. However, these herbicides display no auxin activity of their own [25] nor are they clearly anti-auxins in the sense of a competitive inhibition [3].

Although the inhibition of cell division and cell elongation by most of these herbicides has frequently been observed [9, 26–28], a more detailed study including the auxin-specific responses of the oat coleoptile has been reported only for *chlorfenprop-methyl* [3] and *diclofop-methyl* [25]. Figure 43 describes the inhibitory effects of *chlorfenprop-methyl* and the corresponding acid *chlorfenprop* on *IAA*-transport. Although the ester is more inhibitory, here too the acid should be the active molecule. The possible reason for this descrepancy has been discussed (p. 180). In a study with *chlorfenprop-methyl* [3], the comparative sensitivity of *IAA*-induced responses in oat coleoptiles to *chlorfenprop-methyl* was in the order of *IAA*-transport = *IAA*-dependent elongation > *IAA*-stimulated proton excretion. In the maize coleoptile the order of sensitivity was *NAA*-binding > *NAA*-induced elongation = *NAA*-aspartate formation > *NAA* uptake. However, the maize system may not be relevant for several reasons: (a) the binding interference of the dextrorotatory and the laevorotatory enantiomer did not differ significantly, (b) the sensitivity of other auxin-induced responses was two orders of magnitude lower than in oat and (c) maize is not particularly sensitive to *chlorfenprop-methyl*. The inhibitory actions of *chlorfenprop-methyl*, *diclofop-methyl*, and *flamprop-isopropyl* [9] on the oat coleoptile system, on the other hand, appear to be more specific.

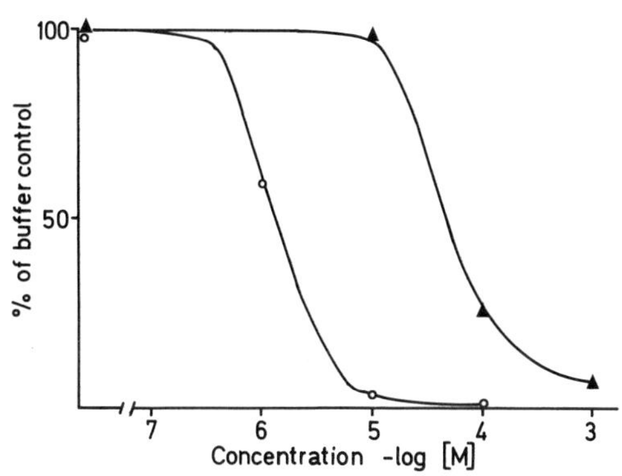

Fig. 43. Inhibition of polar transport of ^{14}C-IAA in oat coleoptile segments by *chlorfenprop-methyl* and *chlorfenprop* [3], ○; *chlorfenprop-methyl*, ▲; *chlorfenprop*

An antagonistic interaction with one or more of the auxin herbicides *2,4-D*, *MCPA*, *2,4-DB*, and *dicamba* has been reported for *benzoylprop-ethyl*, *flamprop-methyl*, *flamprop-isopropyl*, and *diclofop-methyl* [25, 26, 29–33]. The antagonism included not only cell elongation and herbicidal activity, but also uptake, hydrolysis, and complex formation in one case.

In experiments with oat coleotiles, *IAA* was not competitive with *chlorfenprop-methyl* [3], but *IAA* and *dicamba* appeared to be competitive with *diclofop-methyl* [25]. Therefore, it is not certain whether the two herbicides affect identical sites or whether they act at different, but closely related sites. Apparently, a wide diversity exists in the structure-activity relationships of antiauxins and auxin-transport inhibitors and in their relationship to binding and binding interference with auxins [6, 34, 35].

To sum up, it may be reasonably postulated that the site of molecular action of the herbicides listed in Table 54 is located in a lipophilic compartment, which is most likely the plasma membrane. At this location, several possible sites for the binding of the herbicidally active acid may exist, and interference may then occur with the binding of endogenous *IAA*, thereby affecting auxin action. The auxin action primarily affected may be the auxin-transport system, but other auxin-induced responses may be equally inhibited primarily or secondarily.

IV. General Metabolic Effects and Phytotoxicity

Several reports indicate that young, expanding tissues and meristems are most sensitive to these auxin-inhibiting herbicides, and that these tissues must be contacted by the herbicide to ensure satisfactory herbicidal action [27, 28]. Early effects observed include increased membrane permeability [2, 36] and at the subcellular level changes in chloroplast ultrastructure [37, 38] with subsequent damage to other endogenous membrane systems. The cellular endomembrane system, therefore, appears to be particularly sensitive to damage [46], either directly or indirectly, and the consequence at the biochemical level is an "intracellular mix" resulting in autolysis [39]. During this process, the hydrolytic and oxidative enzymes of the lysosomal compartment, i.e. the vacuole, are liberated and subsequently act on the cytoplasmic constituents. Total cellular destruction results. Degradation of polymers, i.e. starch, protein, lipids, nucleic acids, etc., is observed with a transient accumulation of the degradation products [40]. It is therefore not unexpected that general metabolic processes like assimilate distribution [41], lipid synthesis [42], and phosphate uptake, transport, and metabolism [43, 44] are also inhibited. The synthesis of lipids was found to be particularly sensitive after *diclofop-methyl* application; this process may therefore be one of the early processes affected [42].

A direct inhibitory effect on the basic metabolic pathways of photosynthesis, respiration, protein synthesis, and nucleic acid synthesis has been observed only at inhibitor concentrations above 10 μM. Therefore, the significance of these effects as related to the primary herbicidal action is not clear [39, 42, 45]. Many of the observed effects may be secondary effects as discussed in Sect. III.

References for K

1. Beyer, E.M., Johnson, A.L., Sweetser, P.B.: Plant Physiol. *57*, 839 (1976)
2. Schmidt, Th., Fedtke, C., Schmidt, R.R.: Z. Naturforsch. *31c*, 252 (1976)
3. Andreev, G.K., Amrhein, N.: Physiol. Plant. *37*, 175 (1976)
4. Scott, R.M., Sampson, R.M., Jordan, D.: Proc. Brit. Crop Prot. Conf. – Weeds, Vol. 2, p. 723 (1976)
5. Simon, H.: personal communication
6. Katekar, G.F., Geissler, A.E.: Plant Physiol. *66*, 1190 (1980)
7. Hill, B.D., Stobbe, E.H., Jones, B.L.: Weed Res. *18*, 146 (1978)
8. Hill, B.D., Stobbe, E.H.: Weed Res. *18*, 223 (1978)
9. Jeffcoat, B., Harries, W.N.: Pestic. Sci. *6*, 283 (1975)
10. Jeffcoat, B., Harries, W.N., Thomas, D.B.: Pestic. Sci. *8*, 1 (1977)
11. Beynon, K.J. et al.: Pestic. Biochem. Physiol. *4*, 98 (1974)
12. Roberts, T.R.: Pestic. Sci. *8*, 463 (1977)
13. Roberts, T.R.: Pestic. Biochem. Physiol. *7*, 373 (1977)
14. Jeffcoat, B., Harries, W.N.: Pestic. Sci. *4*, 891 (1973)
15. Fedtke, C., Schmidt, R.R.: Weed Res. *17*, 233 (1977)
16. Todd, B.G., Stobbe, E.H.: Weed Sci. *25*, 382 (1977)
17. Shimabukuro, R.H., Walsh, W.C., Hoerauf, R.A.: J. Agric. Food Chem. *27*, 615 (1979)
18. Donald, W.W., Shimabukuro, R.H.: Physiol. Plant. *49*, 459 (1980)
19. Boldt, P.F., Putnam, A.R.: Weed Sci. *28*, 474 (1980)
20. Zacher, H., Hoppe, H.H.: Z. PflKrankh. PflSchutz, Sonderh. IX, 179 (1981)
21. Köcher, H., Lötzsch, K.: Z. PflKrankh. PflSchutz, Sonderh. IX, 171 (1981)
22. Todd, B.G.: Weed Abstr. *29*, 469 (1980)
23. Dusky, J.A., Davis, D.G., Shimabukuro, R.H.: Physiol. Plant. *49*, 151 (1980)
24. Pont, V., Collet, G.F.: Phytochem. *19*, 1361 (1980)
25. Shimabukuro, M.A. et al.: Pestic. Biochem. Physiol. *8*, 199 (1978)
26. Morrison, I.N., Hill, B.D., Dushnicky, L.G.: Weed Res. *19*, 385 (1979)
27. Ott, R., Wimschneider, W.: Z. PflKrankh. PflSchutz, Sonderh. 8, 449 (1977)
28. Coupland, D. et al.: Weed Res. *18*, 123 (1978)
29. Schreiber, M.M., Orwick, P.L.: Abstr. Weed Sci. Soc. Amer. Meet. no. 6 (1977)
30. O'Sullivan, P.A., van den Born, W.H.: Weed Res. *20*, 53 (1980)
31. Todd, B.G., Stobbe, E.H.: Weed Sci. *28*, 371 (1980)
32. Qureshi, F.A., van den Born, W.H.: Weed Sci. *27*, 202 (1979)
33. Fletcher, R.A., Drexler, D.M.: Weed Sci. *28*, 363 (1980)

33a. Hill, B.D., Todd, B.G., Stobbe, E.H.: Weed Sci. *28*, 725 (1980)

34. Katekar, G.F., Geissler, A.E.: Plant Physiol. *60*, 826 (1977)
35. Geissler, A.E., Huppartz, J.L., Katekar, G.F.: Pestic. Sci. *6*, 441 (1975)
36. Crowley, J., Prendeville, G.N.: Canad. J. Plant Sci. *59*, 275 (1979)
37. Davis, D.G., Brezeanu, A.: Canad. J. Bot. *57*, 2006 (1979)
38. Brezeanu, A., Davis, D.G., Shimabukuro, R.H.: Canad. J. Bot. *54*, 2038 (1976)
39. Fedtke, C.: Weed Res. *12*, 325 (1972)
40. Chow, P.N.P., LaBerge, D.E.: J. Agric. Food Chem. *26*, 1134 (1978)
41. Chow, P.N.P. et al.: Canad. J. Plant Sci. *56*, 429 (1976)
42. Hoppe, H.H.: Z. PflKrankh. PflSchutz, Sonderh. IX, 187 (1981)
43. Bergmannova, E., Taimr, L.: Weed Res. *20*, 1 (1980)
44. Bergmannova, E., Taimr, L.: Weed Res. *20*, 243 (1980)
45. Bugg, M.W., Cohen, W.S., Rieck, C.E.: Weed Abstr. *29*, 243 (1980)
46. Albaugh, G.P., Regeler, D.L., Frick, H.: Abstr. Weed Sci. Soc. Amer. Meet. no. 214 (1980)
47. Nestler, H.J., Bieringer, H.: Z. Naturforsch. *35*, 366 (1980)

L. Aromatic Amino Acid Biosynthesis

The inhibition of aromatic amino acid biosynthesis has early been recognized as a possible herbicidal target because this biosynthetic pathway occurs only in microorganisms and plants but not in animals. In a study on the inhibition of shikimate dehydrogenase – the enzyme responsible for the synthesis of the intermediate shikimic acid (Scheme 1) – active inhibitors in vitro were found, but none of these turned out to be herbicidal [1]. It can therefore be regarded as a confirmation of the original concept that the herbicide *glyphosate* (Table 56) is an inhibitor of the enzyme 5-enolpyruvylshikimate-3-phosphate synthase (Scheme 1), only two steps further down from shikimate dehydrogenase in the biosynthetic pathway of the aromatic amino acids. However, the history of the investigation of the *glyphosate* mode of action also illustrates the difficulties of clearly recognizing a primary mode of molecular action which is obscured by secondary, regulatory, and phytotoxic responses.

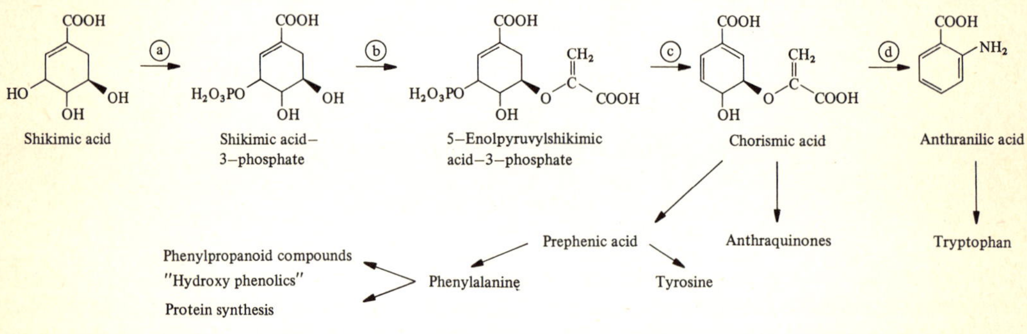

Scheme 1. Reactions occurring in the synthesis and further metabolism of aromatic amino acids which are relevant to the action of *glyphosate*. ⓐ = shikimate kinase, ⓑ = 5-enolpyruvylshikimate-3-phosphate synthase, ⓒ = chorismate synthase, ⓓ = anthranilate synthase. (Adapted from [14]).

I. Primary Mode of Action and Regulatory Responses

The assumption that *glyphosate* (Table 56) is an inhibitor of aromatic amino acid biosynthesis was made very early after experiments with *Lemna* where tyrosine and phenylalanine alleviated the inhibition of growth caused by the herbicide [2]. Unfortunately, however, the alleviating action of phenylalanine and tyrosine is not generally observed, and doubts as to the relevance of aromatic amino acid biosyn-

Table 56. Structures of the herbicide *glyphosate* and the related growth regulator *glyphosine*

Common name	Structure	Water sol. ($g\,l^{-1}$)
Glyphosate	$HOOC-CH_2-NH-CH_2-PO_3H_2$	12
Glyphosine	$HOOC-CH_2-N(CH_2-PO_3H_2)_2$	248

thesis as a primary herbicidal target were then repeatedly expressed [3, 4]. In search of the primarily sensitive enzyme, chorismate mutase, prephenate dehydrogenase, and 3-deoxy-D-arabino-heptulonosate-7-phosphate synthetase have been studied and found to be either insensitive or sensitive at a herbicide concentration of 10 mM [5, 34].

Before further discussing the effects of *glyphosate*, it should be remembered that aromatic amino acids fulfill two equally important roles in plant metabolism: (a) forming protein amino acids and (b) serving as starting compounds for the phenylpropanoid metabolism. The latter role is of extreme quantitative importance in the case of phenylalanine, since the corresponding phenylpropanoid compounds are "mass products" formed during the growth of the plant. They include, e.g., many pigments and the cell wall polymer lignin:

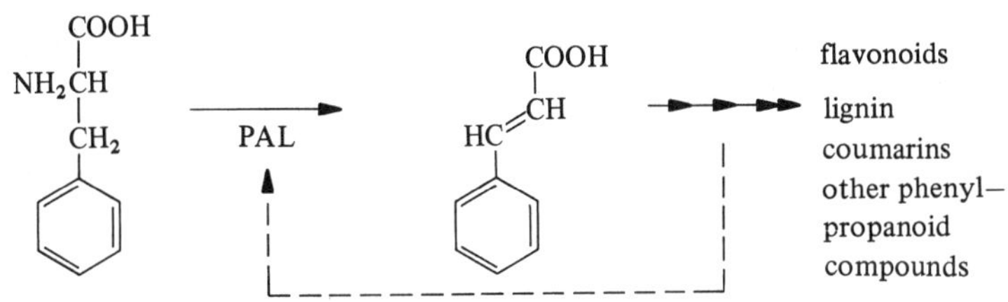

phenylalanine trans–cinnamic acid

The key enzyme of phenylpropanoid metabolism, phenylalanine ammonia lyase (PAL), is thought to be subject to end product repression and thereby to regulate the whole pathway [6]. Accordingly, extractable PAL-activity is often negatively correlated with the pool sizes of the aromatic amino acids phenylalanine and tyrosine.

The finding that *glyphosate* strongly increases the activity of extractable PAL is of great relevance for the understanding of its mode of action. This finding may be interpreted by two fundamentally different hypotheses: (a) the increase could be caused by a more or less direct herbicidal influence and would then initiate an increased rate of phenylpropanoid synthesis or (b) the increase could be caused by indirect herbicidal interference and reflect the release of PAL from the natural end product repression, i.e. it could be caused by a decreased rate of phenylpropanoid synthesis:

Table 57. Influence of *glyphosate*, L-α-aminooxy-β-phenylpropionic acid (AOPP) and combinations thereof on phenylpropanoid metabolism (4); ↑ increase; ± no change; ↓ decrease

Additions	Phenylalanine ammonia-lyase		Phenyl-alanine amount	„Soluble OH-phenolics"	
	Enzyme potential (extractable activity)	Enzyme activ. (in vivo)		Per organ	Per fresh weight
Glyphosate	↑	↑	↓*	↓	±↑
AOPP	↑	↓	↑	↓	↓
Glyphosate + AOPP	↑	(±)	±	↓	±

* Only in roots and aquatic plants

(a) A higher rate of phenylpropanoid synthesis could lead to the accumulation of growth-inhibiting levels of "hydroxyphenolics" and thereby produce the herbicidal effects of *glyphosate* [3, 7]. However, when this hypothesis was checked by measuring the "soluble hydroxyphenolics" after application of *glyphosate* and of the PAL-inhibitor L-α-aminooxy-β-phenylpropionic acid (AOPP), individually and in combination (Table 57), no clear confirmation was obtained [4, 4a]. Although *glyphosate* and AOPP displayed opposite effects with respect to in vivo PAL-activity and phenylalanine concentrations, an accumulation of growth-inhibiting levels of "soluble hydroxyphenolics" was never observed.

(b) A natural regulatory response of PAL-activity would require a decreased rate of phenylpropanoid synthesis, which has actually been found [4, 8]. However, two findings are inconsistent with an effective inhibitor of phenylalanine synthesis and complicate the corresponding explanation: the hydroxyphenolics are only decreased on a per plant or per organ basis (Table 57), and the levels of free aromatic amino acids are not decreased very strongly or often not at all [4, 9]. A possible explanation for these contradictory results might come from a consideration of phenylalanine liberation from natural protein turnover or from phenylpropanoid recycling in plant metabolism. The extent of the latter will, however, be very low. Also, subcellular compartmentalization might contribute to an understanding of the contradictory results. In isolated soybean leaf cells a clear antagonism between *glyphosate* and aromatic amino acids was observed [10], and in wheat a strong reduction of aromatic amino acid pools could be obtained after *glyphosate* application [11].

Recently, it has been reported that *glyphosate* is a very effective inhibitor of 5-enolpyruvylshikimate-3-phosphate synthase, an enzyme of aromatic amino acid biosynthesis (see Scheme 1 on p. 184). The I_{50}-value was ca. 5 μM for the sequence shikimate→anthranilate and ca. 20 μM for the reaction shikimate-3-phosphate→5-enolpyruvylshikimate-3-phosphate under the respective standard conditions in an extract from an auxotroph mutant of the bacterium *Aerobacter aerogenes* [12–14]. The relevance of these in vitro data for the higher plant in vivo can be concluded from (a) the inhibition of ^{14}C-shikimate incorporation into phenylalanine in buckwheat [8] and (b) the accumulation of shikimate in plants treated with *glyphosate* [15]. The fact that shikimate and not shikimate-3-phosphate accumulates in the treated plants has been attributed to the presence of a phosphatase.

It might be speculated that *glyphosate* acts as an anti-metabolite of 2-phosphoenol pyruvate (PEP) which is condensed with shikimate-3-phosphate in the 5-enolpyruvyl-shikimate-3-phosphate synthase reaction. However, although *glyphosate* inhibits competitively with respect to PEP but not with respect to shikimate-3-phosphate, it cannot be considered as a general PEP-antagonist since other reactions in which PEP participates are insensitive to *glyphosate* [35].

II. Interference of Glyphosate and Glyphosine with Other Metabolic Reactions

Whereas *glyphosate* is used as a herbicide, the related compound *glyphosine* is used to increase the sugar content in cane (Table 56). However, *glyphosine* may well be converted into *glyphosate* in the plant and exert its effect via inhibition of aromatic amino acid biosynthesis. *Glyphosine* is also known to induce chlorosis [16]. Similar to the chlorosis-inducing herbicides (p. 99), *glyphosine* inhibits the formation of chloroplast ribosomes, chloroplast RNA, grana stacking, and pigment formation. *Glyphosate* may also reduce carotenoid and chlorophyll contents in treated tissue [17]. At a concentration of 5 mM, *glyphosate* does not inhibit photosynthesis in pea chloroplasts [18]. The inhibition of O_2-evolution in *Scenedesmus* at 0.7 mM may occur indirectly since it increased with time [19]. Similarly, respiration is not inhibited or uncoupled at μM concentrations, whereas mM concentrations do inhibit respiration in some cases [20]. After prolonged herbicidal action many different cellular biosynthetic processes are affected, e.g. protein synthesis [10, 21], photosynthesis, and respiration [22]. Simultaneously, the translocation of assimilates is reduced. Another possibly indirect effect is the increased production of ethylene and the increase of cellulase activities in the abscission zone of bean seedlings [23].

Only a few hours after *glyphosate* application is the transpiration rate strongly affected [24, 25]. Since this inhibition of transpiration may be reduced by the addition of phenylalanine and tyrosine, the authors explained this by an inhibited protein synthesis in the treated tissue [26]. However, since the inhibition of protein synthesis is already an indirect effect and is accompanied by other indirect *glyphosate* effects, any of these might affect stomatal opening and transpiration [25].

Effects of glyphosate on mitotic chromosomes of rye have also been reported. At 0.5 lb/acre *glyphosate* induces a substantial amount of clumping of bivalents and unsynchronized divisions [27] whereas at 50 mgl^{-1} it prevents *Eichhornia* interphase cells from entering mitosis [27a].

III. Phytotoxicity from Glyphosate

A peculiarity of *glyphosate* action is the slow herbicidal action in the field. This may be due to the fact that the pools of free and bound aromatic amino acids and the pools of phenylpropanoid compounds are large enough to allow a considerable amount of growth before they limit further growth. In germinating seeds, these pools are much smaller and the herbicidally induced reduction of the growth rate obviously occurs more rapidly. In the field, *glyphosate* is applied to the established plants where the herbicide moves freely up- and down-stream. No pre-emergence activity can be observed.

From the available information it is assumed that the inhibition of aromatic amino acid biosynthesis represents the primary molecular mode of *glyphosate* action, and that the other observed effects and inhibitions are consequences therefrom. However, the possibility of additional molecular sites of *glyphosate* action, at μM or slightly higher concentrations, cannot at present be completely excluded.

In the presence of *glyphosate* the seedling germinates normally but the subsequent growth is slowed down increasingly until it stops completely. The plant then decays [28, 29]. One early effect observed in bean plants and isolated cells was an inhibition of ion absorption [29]. The reason for the tolerance observed in *Euphorbia esula* is not entirely clear [30]. On an ultrastructural level, disruption of the chloroplast envelope, swelling of the endoplasmatic reticulum, and a progressive destruction have been observed [31, 32]. However, it is at present not possible to ascribe the final phytotoxic effects to one single phytotoxic molecule. One possibility is ammonia toxicity, since ammonia is no longer consumed in the aromatic amino acid biosynthesis and may therefore accumulate [33]. Eventually, numerous metabolic disfunctions will add to the complete loss of growth and metabolic integrity.

References for L

1. Baillie, A.C. et al.: Pestic. Sci. *3*, 113 (1972)
2. Jaworski, E.G.: J. Agric. Food Chem. *20*, 1195 (1972)
3. Duke, S.O. Hoagland, R.E.: Plant Sci. Lett. *11*, 185 (1978)
4. Duke, S.O., Hoagland, R.E., Elmore, C.D.: Plant Physiol. *65*, 17 (1980)
4a. Hoagland, R.E.: Weed Sci. *28*, 393 (1980)
5. Roisch, U., Lingens, F.: Angew. Chemie *86*, 408 (1974)
6. Amrhein, N., Gerhardt, J.: Biochim. Biophys. Acta *583*, 434 (1979)
7. Hoagland, R.E., Duke, S.O., Elmore, C.D.: Plant Sci. Lett. *13*, 291 (1978)
8. Holländer, H., Amrhein, N.: Plant Physiol. *66*, 8 (1980)
9. Haderlie, L.C., Widholm, J.M., Slife, F.M.: Plant Physiol. *60*, 40 (1977)
10. Tymonko, J.M., Foy, C.L.: Plant Physiol. Suppl. *61*, 41 (1978)
11. Nilsson, G.: Weed Abstr. *27*, 407 (1978)
12. Amrhein, N., Schab, J., Steinrücken, H.C.: Naturwiss. *67*, 356 (1980)
13. Amrhein, N. et al.: Plant Physiol. *66*, 830 (1980)
14. Steinrücken, H.C., Amrhein, N.: Biochem. Biophys. Res. Comm. *94*, 1207 (1980)
15. Amrhein, N.: Plant Physiol. *66*, 830 (1980)
16. Croft, S.M. et al.: Biochim. Biophys. Acta *335*, 211 (1974)
17. Abu-Irmaileh, B.E., Jordan, L.S.: Weed Sci. *26*, 700 (1978)
18. Richard, E.P., Goss, J.R., Arntzen, C.J.: Weed Sci. *27*, 684 (1979)
19. van Rensen, J.J.S.: Weed Abstr. *25*, 148 (1976)
20. Hanson, C.L., Rieck, C.E.: Weed Abstr. *25*, 194 (1976)
21. Cole, D.J., Dodge, A.D.: Plant Physiol. Suppl. *63*, 96 (1979)
22. Sprankle, P., Meggitt, W.F., Penner, D.: Weed Sci. *23*, 235 (1975)
23. Abu-Irmaileh, B.E., Jordan, L.S., Kumamoto, J.: Weed Sci. *27*, 103 (1979)
24. Shaner, D.L.: Weed Sci. *26*, 513 (1978)
25. Shaner, D.L., Lyon, J.L.: Weed Sci. *28*, 31 (1980)
26. Shaner, D.L., Lyon, J.L.: Abstr. Weed Sci. Soc. Amer. Meet. no. 200 (1980)
27. Boyle, W.S., Evans, J.O.: J. of Heredity *65*, 250 (1974)
27a. Goltenboth, F.: Weed Abstr. *29*, 427 (1980)

28. Hoagland, R.E.: Plant Physiol. Suppl. *59*, 78 (1977)
29. Brecke, B.J., Duke, W.B.: Plant Physiol. *66*, 656 (1980)
30. Gottrup, O. et al.: Weed Res. *16*, 197 (1976)
31. Campell, W.F., Evans, J.O., Reed, S.C.: Weed Sci. *24*, 22 (1976)
32. Uotila, M., Evjen, K., Iversen, T.-H.: Weed Res. *20*, 153 (1980)
33. Killmer, J.L., Widholm, J.M., Slife, F.W.: Abstr. Proc. North Central Weed Sci. Soc. Amer. Conf. *34*, 4 (1979)
34. Roisch, U., Lingens, F.: Hoppe-Seyler's Z. Physiol. Chem. *361*, 1049 (1980)
35. Amrhein, N.: personal communication

M. Other Herbicides and Mechanisms

I. Competition with Nitrate

The total herbicide *sodium chlorate* may be reduced by the enzyme nitrate reductase to yield the highly reactive and acutely phytotoxic chlorite ion (ClO_2^-):

$$\begin{matrix} NO_3^- \\ ClO_3^- \end{matrix} \xrightarrow{\text{nitrate reductase, NADH}} \begin{matrix} NO_2^- \\ ClO_2^- \end{matrix}$$

This has been shown to be true for *Chlorella* [1–3], *Chlamydomonas* [3], and tomato [4]. The same obviously applies to other higher plants which have been used as starting material for the selection of resistant lines, as discussed below. However, ClO_3^- does not act as an inducer of nitrate reductase. The alternative reduction of nitrate or clorate by nitrate reductase is a competitive process.The k_m for chlorate has been estimated as 2.3 mM for *Chlamydomonas reinhardii*, 1.2 mM for *Chlorella vulgaris*, 3.9 mM for *Chlorella pyrenoidosa*, and 4 mM for tomato. The reduction product chlorite leads to the rapid inactivation of nitrate reductase and possibly other enzymes in the vicinity and may thereby display its phytotoxic action.

The conclusion that activation of clorate by reduction to chlorite via nitrate reductase constitutes the mode of the herbicidal action for *sodium chlorate* is supported by the following observations:

1. tomato plants are only sensitive to the herbicidal action of chlorate when grown either on nitrate or in the absence of a nitrogen source, but not when grown on ammonia,
2. high nitrate concentrations tend to counteract the phytotoxicity caused by chlorate,
3. urea counteracts the phytotoxicity of chlorate in wheat plants, and
4. the toxicity is highest under conditions of high in vivo-nitrate reductase activity and vice versa (e.g. in the light or in the dark).

In conclusion, in vivo-nitrate reductase activity and the concentration ratio of free ClO_3^- to NO_3^- are the important factors for the occurrence of phytotoxicity after chlorate application.

The presence of chlorate in growth or culture media has been used for the selection of resistant plants which lack the nitrate reduction pathway. Resistant mutants have been obtained with *Arabidopsis thaliana* [5, 6] and *Nicotiana tabacum* [7], and with *Rosa damascena* cell cultures [8]. Although nitrate reductase deficiency is a common mechanism of resistance involving either loss of the Mo-cofactor or loss of enzyme synthesis [7], competition for nitrate uptake [5, 6] and loss of catalase activity upon chlorate addition appear to represent additional resistance mechanisms.

II. Mefluidide

The mode of action of the growth regulator and post-emergence herbicide *mefluidide* is unknown. The influence of *mefluidide* on several metabolic pathways, including respiration, photosynthesis, and protein synthesis, have been studied, but only slight inhibitions have been observed [9, 10].

Table 58. Structures of the herbicides *sodium chlorate* and *mefluidide*

Common name	Structure	Water sol.
Sodium chlorate	$NaClO_3$	$790\ g\ l^{-1}$
Mefluidide	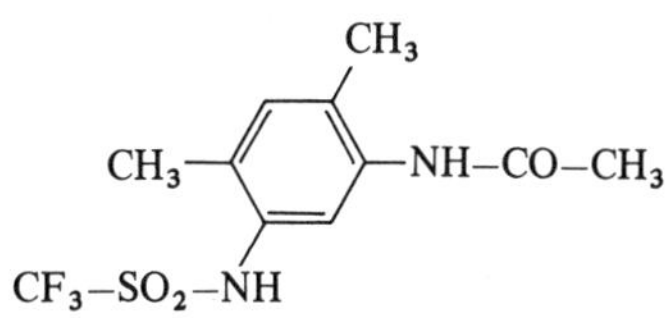	$180\ mg\ l^{-1}$

References for M

1. Solomonsen, L.P., Vennesland, B.: Plant Physiol. *50*, 421 (1972)
2. Tromballa, H.W., Broda, E.: Arch. Microbiol. *78*, 214 (1971)
3. Rhodes, D., Filner, P.: Plant Physiol. Suppl. *68*, no. 127 (1979)
4. Hofstra, J.J.: Physiol. Plant. *41*, 65 (1977)
5. Doddema, H., Telkamp, G.P.: Physiol. Plant. *45*, 332 (1979)
6. Doddema, H., Telkamp, G.P.: Physiol. Plant. *45*, 339 (1979)
7. Müller, A., Grafe, R.: Molec. gen. Genet. *161*, 67 (1978)
8. Imbrie, C.W., Murphy, T.M.: Plant Physiol. Suppl. *65*, no. 200 (1980)
9. Truelove, B., Davis, D.E., Pillai, C.G.P.: Weed Sci. *25*, 360 (1977)
10. Bugg, M.W., Hayes, R.M.: Abstr. 31st Annu. Meet. Southern Weed Sci. Soc. Amer. 245 (1978)

Subject Index

Insektizide, Bakterizide. Oomyceten-Fungizide. Biochemische und biologische Methoden. Naturstoffe/ Insecticides. Bactericides. Oomycete Fungicides. Biochemical and Biological Methods. Natural Products

1981. 105 figures, 92 schemes. XVI, 512 pages (191 pages in German)
(Chemie der Pflanzenschutz- und Schädlingsbekämpfungsmittel, Band 6)
ISBN 3-540-10307-4

Contents:
J. A. A. Renwick, J. P. Vité: Biology of Pheromones. – *H. J. Bestman, O. Vostrowsky:* Chemistry of Insect Pheromones. – *M. Boneß:* Die praktische Verwendung von Insektenpheromonen. – *J. P. Edwards, J. J. Menn:* The Use of Juvenoids in Insect Pest Management. – *K. Bauer, D. Berg, E. Bischoff, H. v. Hugo, P. Kraus:* Pflanzenschutzpräparate mikrobieller Herkunft. – *C. Fest:* Insektizide Phosphorsäureester. – *T. Egli, E. Sturm:* Bacterial Plant Diseases and Their Control. – *T. H. Staub, A. Hubele:* Recent Advances in the Chemical Control of Oomycetes. – *W. Maas, R. v. Hes, A. C. Grosscurt, D. H. Deul:* Benzoylphenylurea Insecticides. – *H. H. Cramer, B. Zeller:* Zur Problematik des biologischen Pflanzenschutzes in der Landwirtschaft.

Springer-Verlag
Berlin
Heidelberg
New York

Spezielle Chemie der Herbizide. Anwendung und Wirkungsweise/ Special Chemistry of Herbicides. Applications and Mechanisms

1982. VII, 455 Seiten (40 Seiten in Englisch)
(Chemie der Pflanzenschutz- und Schädlingsbekämpfungsmittel, Band 8)
ISBN 3-540-10778-9

Contents/Information:
Seit im Jahr 1977 Band 5 des Werkes zum Thema „Herbizide" erschien, gewannen die Herbizide erneut an Bedeutung. Neben der Einführung von Versuchs- und Handelsprodukten, die Substanzgruppen entstammen, deren Bedeutung schon Ende 1976 bekannt war, wurden auch spektakuläre Entdeckungen gemacht. Herausgeber und Verlag haben sich daher entschlossen, diesen Band den Neuentwicklungen auf dem Gebiet der Herbizide zu widmen.
Schon 1976 bekannte Verbindungsklassen, wie die der Diarylether-oxyalkancarbonsäuren, ergaben eine Fülle neuer Herbizide mit meist selektiver Gräserwirkung im Getreide. Mitarbeiter der Firma Hoechst AG haben daher für diesen Band einen entsprechenden Beitrag zur Verfügung gestellt.
H. J. Nestler: **Phenoxy-phenoxy-Propionsäure-Derivate und verwandte Verbindungen (Phenoxy-phenoxy propionic Acid Derivatives and Related Compounds)**
Das immer noch aktuelle Problem der Wildhaferbekämpfung wird in einem Beitrag der Shell Int. Res. behandelt, eine Firma mit speziellen Erfolgen bei Flughafer-Herbiziden.
E. Haddock, R. G. Turner: **Wild Oat Herbicides (Wildhafer-Herbizide)**
Der Herausgeber selbst hat die Herbizid-Fortschritte 1976 bis 1980 bearbeitet, einschließlich einer vollständigen Patentübersicht.
R. Wegler, L. Eue: **Neue Herbizide (New Herbicides)**
Erstmals werden neben Patenten bzw. Offenlegungsschriften der Bundesrepublik Deutschland die an Bedeutung rasch zunehmenden Europa-Patente vollständig aufgeführt. In Anbetracht der wachsenden Bedeutung Japans als Erfinder, Hersteller und Verbraucher von Pflanzenschutzmitteln, wurden japanische Patente mit besonderer Sorgfalt erfaßt. Zusammen mit den Patentergänzung aus Band 5 liegt nun ein fast lückenloser Überblick über die wichtigsten Patente vor. Aus den Patentergänzungen lassen sich wichtige Hinweise auf Neuentwicklungen ablesen, was für eine ökonomische Forschungsplanung von großer Bedeutung ist.